Math
Study Skills

ALAN BASS

San Diego Mesa College

PEARSON

Addison
Wesley

Boston   San Francisco   New York
London   Toronto   Sydney   Tokyo   Singapore   Madrid
Mexico City   Munich   Paris   Cape Town   Hong Kong   Montreal

Publisher: Greg Tobin

Editor in Chief: Maureen O'Connor

Executive Acquisitions Editor: Jennifer Crum

Editorial Assistant: Antonio Arvelo

Senior Managing Editor: Karen Wernholm

Senior Production Supervisor: Tracy Patruno

Senior Designer: Barbara T. Atkinson

Marketing Manager: Jay Jenkins

Manufacturing Manager: Evelyn Beaton

Cover Design, Text Design, and Composition: Sandy Silva

Photos on pages 65, 66, and 69 are courtesy of the author.

Library of Congress Cataloging-in-Publication Data
Bass, Alan, 1973-
     Math study skills / Alan Bass. -- 1st ed.
        p.   cm.
     Includes bibliographical references and index.
     ISBN 0-321-51307-X (alk. paper)
     1. Mathematics--Study and teaching.   2. Mathematics--Problems,
        exercises, etc.   3. Study Skills.   I. Title.

QA11.2.B364   2007
510.71--dc22

                                                    20077019697

ISBN-10: 0-321-51307-X
ISBN-13: 978-0-321-51307-6

3 4 5 6 7 8 9 10—BB—11 10 09 08 07

Contents

Preface v

1 Introduction 1

2 What Makes Math Different 9

3 Learning Styles 19

4 Math Anxiety 25

5 Managing Your Time 32

6 Your Class Notebook 40

7 Your Textbook and Homework 47

8 Class Time and Note Taking 54

9 Retention and General Study Strategies 63

10 Test Taking 75

Math Study Skills for College Students is an honest and considered look at the issue of student success addressed directly and specifically to college math students, particularly those at the developmental level. In developing the book, we focused entirely on addressing what college students need to know and do to overcome the challenge of passing a math class. We made it a point to exclude all concepts not directly relevant to student success. Here are some elements that we feel make this book a helpful tool for students:

▮▮▮ Strong Focus on Relevance and Results

Unlike a lot of study skills materials, *Math Study Skills for College Students* focuses on the skills and techniques needed to be successful in math, not on the psychology behind them. We have listened carefully to college students about what they need to succeed and have only included information deemed relevant by students.

▮▮▮ Short and Sweet

We made a disciplined effort to keep the book at a length that makes it easy to incorporate into a math class. Yet, there is enough substance for it to be used as the main text in a Study Skills course.

▮▮▮ Interactive and Engaging

The writing style is casual and very engaging. The author often asks questions for the reader to think about while reviewing a topic. In addition, exercises are included throughout the text to help students immediately apply the skills they are learning and discover what approach will be most beneficial for them.

▮▮▮ Strong Time Management Content

There is a whole chapter devoted to improving time management skills. In this chapter, we explore the demanding expectations placed on students at the college level and provide exercises that help students understand the College Rule of time management.

▌▌▌ Understanding and Honest

This book makes no effort to sugarcoat any important issues that are better addressed directly and honestly. At the same time, it relates to students' concerns, includes an entire chapter on overcoming math anxiety, and shows students how to stay positive and change their circumstances for the better.

ACKNOWLEDGEMENTS

This book would never have been possible without the work of some outstanding colleagues. I feel more like a medium than an author. The professors involved in the Pathways Through Algebra project have been absolutely instrumental in developing this material. First and foremost, my mentor Terrie Teegarden at San Diego Mesa College started the Pathways project and took me on in spite of my being so green. Thank you so much for everything you do Terrie, for me and for our students. Noelle Eckley of Lassen College and Jennifer McCandless of Shasta College enthusiastically piloted the first draft as part of the Pathways project and provided wonderful feedback. In addition, the following professors piloted the second draft: Allison Damoose, Michelle "Toni" Parsons, Joan Adaskin, and Holly Bass. Holly is my wife; anything good that comes from me is only a reflection of her influence in my life (seriously). Addison-Wesley has been an absolute pleasure to work with. In particular, thanks to Jennifer Crum for believing in the book and brainstorming with me. Thanks to Antonio Arvelo for being so on top of issues that came up, and thanks to Tracy Patruno for coordinating superb production and design. Finally, a special thanks to all my students for being honest about their challenges with succeeding in math and about whether my material was useful to them.

And to the students who use this book: You can do it! Grab the mathematical bull by the horns and take what you deserve. This book is dedicated to you.

Alan Bass

1 Introduction

THE PROBLEM

Let's be honest and get straight to the point. Success in mathematics is one of the biggest challenges facing college students today. Only about half of college students who enter developmental level classes (Prealgebra, Beginning Algebra, or Intermediate Algebra) come out with a passing grade. The other half become frustrated and withdraw from the class or, worse, keep coming to class knowing that they have conceded to failure. This happens to many students even though professors consistently report that nearly all students have the potential to pass.

Students of all kinds face this problem for a variety of reasons. Here are examples of typical struggling math students that professors commonly see:

- Darren is good in other subjects and genuinely tries at math but can't seem to make it work. He may blame the instructor or even the school, since he is working hard and feels he is doing his part. Or he may feel like he is just not smart.

- Sean has hated math literally since the day his fifth grade teacher put a fraction on the board. He has seen it all before. He didn't get it then and is burned out now after having these concepts pounded into his head one more time. While Sean is generally motivated, he has a very difficult time staying motivated in his math class. Because of this, his instructor views him as a slacker.

- Bethany has not been in school for some time and is taking math as a review to advance to higher level classes. She is overwhelmed by the pace set in college level classes. She only needs to complete two math classes to earn her degree, but the task seems daunting given the pace and level of expectation. On top of this, her struggling is resulting in math anxiety, which makes it much harder to succeed.

- Helen comes to class and, for the most part, understands the lectures. She takes good notes and understands most of what her professor explains. She sits down to do homework and can do almost every problem. In fact, her homework grades are excellent. But then, when she takes the exams she gets very anxious, freezes up, and performs poorly. She is in danger of failing the class because of her low test scores.

These are all real students I've had. And I have more like them every semester. You may see yourself in parts of these descriptions. If so, you have come the right place.

THE SOLUTION

This book is about how you, the college student, can take immediate control over your circumstances and succeed in your math course. If you struggle in math, the situation is not as bleak as it may seem. Moreover, the solution to the problem is well within reach of the average student. Here are the main points that we will discuss in more detail in this section:

- Poor performance in math is rarely due to a lack of intelligence.

- The key to success in math is not having high intelligence.
  It's having the right approach to studying and learning.

- Math anxiety is a learned habit pattern that can be changed.

Math is, to turn an adage on itself, "a sheep in wolf's clothing." That is, despite our difficulties with it, it is not at all impossible to master if you have the proper tools. And when I speak of "the proper tools" I am not referring to inflated intelligence. Poor performance in math is not due to a lack of intelligence. This statement runs contrary to how most struggling math students view themselves. In fact, when most students hear me say, "Poor performance in math is not due to a lack of intelligence," I can see the unspoken reaction on their faces: "Get real," or "That may apply to some, but not me." What they don't understand is that this assertion is not wishful speculation or positive psycho-babble. It is a statistical fact. Human beings are easily intelligent enough to understand the basic principles of math. I cannot repeat this enough, but I'll try. . . **Poor performance in math is not due to a lack of intelligence!**

The fact is that the key to success in math is in taking an intelligent approach. Students come to me and say, "I'm just not good at math." Then I ask to see their class notebooks and they show me a chaotic mess of papers jammed into a folder. In history class it may be inconsequential to be so unorganized. But in math, it's a lot like taking apart your car's engine, dumping the heap of disconnected parts back under the hood, and then going to a mechanic to ask why it won't run. Students come to me and say, "I'm just not good at math." Then I ask them about their study habits and they say, "I have to do all my studying on the weekend." In math, that's a lot like a trying to do all of your eating or sleeping on the weekends and wondering why you're so tired and hungry all the time. **How you approach math is much more important than how smart you are.** Not understanding this can make failure unavoidable.

Failure in spite of disciplined and well-meaning efforts can make students feel like math is just too hard to handle, so they begin to feel anxious about trying and failing again, and again, and again. This anxiety over math (math anxiety) is a learned behavior in reaction to an unfavorable event, which leads us to the next important point: **Math anxiety is a learned behavior that can be changed.** Did you know that in other industrialized nations, students do not have a fear of math? Our angst with math is not a human phenomenon, it is a cultural phenomenon. And there are concrete steps that you can take to alleviate the problem.

This book will help anyone who has the patience and discipline to apply the principles and techniques contained in it. It will provide ideas and techniques that will help motivate and train you to move from being a struggling math student to being a successful math student!

To make sure that these ideas sink in and to give you a chance to reflect and explore on what will be best for you, I have placed exercises throughout the material for you to do. You can write in the space provided or write on your own paper, but don't skip the exercises! They provide a good way for you to retain and process what you are learning about study skills. All the exercises require honesty and openness about your situation. Here's the first one.

Your Current Study Habits

Write a response using the following statements and questions as prompts. Write your response in the space provided below each prompt.

Notes on what to include:
- How often do you study and for how long?
- How do you begin a study session?
- How do you end a study session?
- How do you prepare for exams?
- Is there anything else that comes to mind?

"My approach to homework and studying in math class is. . . "

"The things I do that I know are effective are. . . "

"The things that I don't do that I know I should do are. . . "

OVERVIEW OF TOPICS

So what is the right approach? How can we use our intelligence in a more intelligent way? And how can we change our thinking so that math is seen as the useful tool that it is and not a class to be suffered through? To answer these questions, let's preview the topics we will discuss in the different chapters of this book without going into detail about the topics.

What Makes Math Different

There are understandable reasons why we become so bogged down with math. There are certain aspects that make it different from other subjects. It requires its own approach. We are going to look at what makes math different. We will see how these differences have confused us into thinking math is more difficult than it actually is.

Learning Styles

Everyone learns best in his or her own way. We are going to find out what style of learning appeals to you and discuss some specific ways of studying math that will turn that learning style into one of your biggest strengths.

Math Anxiety

In this chapter we will explore why math anxiety occurs. We will also look at symptoms of math anxiety. Most importantly, we will discuss specific ways you can crush math anxiety beneath your triumphant heel and become the confident, successful student you deserve to be!

Managing Your Time

Success is going to require change at some level. For many that change is going to be in how you manage that most precious of resources: time. It may be that you need to spend more time working on math. It may be that you need to change how you spend your time. In either case, this chapter will provide tips and insight on time management.

Your Class Notebook

Every semester I see students' academic careers change radically for the better with the use of an organized, thorough course notebook. We will discuss how and why a good notebook is an essential part of your success in a math course. We will also discuss what your class notebook *must* include, what it *should* include, what it *should not* include.

Your Textbook and Your Homework

Have you ever had a class where you bought the textbook and just became annoyed at all those pages in between the homework? Most students have a really hard time getting good use out of a math text. You may be surprised at what a gold mine your textbook is if you know how to approach it. We'll discuss how to deal with its complexity in a way that will maximize its benefit. We'll also give you strategies for making homework much easier and more helpful.

Class Time and Note Taking

Math as an academic subject is actually supposed to make you feel smart, particularly when you're in class. We are going to look at ways to make class time and note taking easier and more effective.

Retention and General Study Strategies

We'll look at certain techniques that are tried and true for memorizing important concepts and formulas. This section will also show you ways to help break the monotony that is often associated with studying math by mixing things up a little.

Test Taking

You probably know that homework is the most important thing you will do in math class. But did you know that homework can become a dangerous crutch that stops you from succeeding on exams? Did you know that you should always take peppermint to a test? Did you know that you probably already have a sample of every test you'll take in your math class? Preparing for and taking tests is as much a skill as public speaking or hitting a golf ball. We're going to turn ourselves into real pros.

HOW TO USE THIS BOOK

In this book you will find a wide variety of information and strategies for being more successful in math class. It is likely that some (or even most) of the skills discussed in this book are things you are well aware of and are already applying in some capacity. Take from this book only what you need. If you already have good study habits in certain areas, keep using them. At the same time, you should give all the ideas a fair chance by reading them through and doing the exercises; then you can judge for yourself whether they will be useful to you. Also keep in mind that if you are not succeeding in math that means something needs to change. The ideas here will make you a better math student, if you act on them—guaranteed!

GENERAL PRINCIPLES OF SUCCESS

In researching for this book, and in trying to achieve goals in my own life, I have found a set of basic principles for success. These principles come up in one form or another in every book on the topic of success, particularly in areas of life that present a major challenge. Since this is a book on how to *succeed*, I felt it would be appropriate to include a summary of these principles.

Principles of Success Are Self-Evident

Information on how to succeed in any area of your life is basically a collection of things that you already know intuitively. Steven Covey, a success researcher and writer, says that the study of how to succeed is "an in-depth study of the obvious." You may read some of the information in this book and say, "I knew that." The question is not, "Did you know that?" The question is, "Are you *using* it?" This brings us to the next point.

Success Requires Change and Action, Not Just Knowledge

There was this sick guy who went to the doctor. He was so impressed with the doctor that he was quite happy and convinced that his ailment would get better. What excited him most was that the doctor gave him a prescription to help solve his problem: "My doctor is so intelligent! He has told me the source of my sickness and given me this wonderful prescription to remedy it." The man went home very excited and sat on his couch and began reading over and over again: "One pill in the morning, one pill in the afternoon, one pill in the evening. One pill in the morning, one pill in the afternoon, one pill in the evening. One pill in the morning, one pill in the afternoon, one pill in the evening." And all along, to his surprise and dismay, he didn't get any better.

The moral is that *reading* a prescription doesn't help. You have to *take* the prescription. In the case of your math class, your "pill" is going to be a system of studying that will make you work harder and more efficiently. To get more to the point: You can read this entire book and do every exercise, but if you don't act on the information it will not do you any good at all. You must be willing to put in the time and effort to see it through.

> ❝ *It is one of the strange ironies of this strange life that those who work the hardest, who subject themselves to the strictest discipline, who give up certain pleasurable things in order to achieve a goal, are the happiest people. When you see 20 or 30 people line up to run a marathon, do you pity them? Better envy them instead.* ❞
>
> —BRUTUS HAMILTON, *National Track and Field Hall of Fame*

Where Should You Focus?

After reading the overview of topics, in what areas do you think you stand to make the most gains in your math class? In the space below, explore three areas of study skills that you think you need to improve on and why it will help you to work on those areas.

2 What Makes Math Different

WHY MATH SEEMS SO TOUGH

There are some good reasons why college math can seem overwhelming if you don't have the right tools for success. Exploring these issues in depth will make them seem less of a challenge and more of a way to actually make math easier. In the process of discussing the pitfalls you face, we will also explore a few things you can do right away to help you get started down the road to success. But before we do, let's see what you think makes math so challenging.

Why Is Math So Tough?

Write a paragraph on what you think makes math a difficult subject to master. Getting these thoughts down on paper is an important step. It can help you understand what your obstacles are and how to overcome them. Bring up as many points as you can think of. Don't hold back! Be honest and frank. So. . .

What makes math so tough?

There are several aspects of what makes math tough for students that I want to address. As mentioned, I will give you ways of dealing with these issues as we explore them.

Math as a Foreign Language

Have you ever said or heard somebody say, "When my math professor is lecturing it's like he's speaking a foreign language"? The student is incorrect on one very important point:

> Math is not LIKE a foreign language, math IS a foreign language.

Consider the following list of math vocabulary words and note which of them have a totally different meaning in English.

- product
- term
- argument
- composition

- quotient
- function
- exponent
- formula

- difference
- domain
- inverse
- factor

And, as you are well aware, math isn't just about the vocabulary terms. There dozens of symbols that a person must be familiar with.

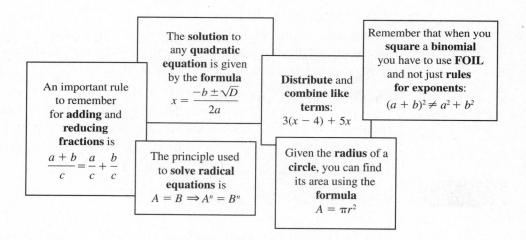

An important rule to remember for **adding** and **reducing fractions** is
$$\frac{a + b}{c} = \frac{a}{c} + \frac{b}{c}$$

The **solution** to any **quadratic equation** is given by the **formula**
$$x = \frac{-b \pm \sqrt{D}}{2a}$$

The principle used to **solve radical equations** is
$$A = B \Rightarrow A^n = B^n$$

Distribute and **combine like terms**:
$$3(x - 4) + 5x$$

Given the **radius** of a **circle**, you can find its area using the **formula**
$$A = \pi r^2$$

Remember that when you **square** a **binomial** you have to use **FOIL** and not just **rules for exponents**:
$$(a + b)^2 \neq a^2 + b^2$$

Math, like chemistry or physics, communicates using a mixture of both English and symbols. But knowing that math is a foreign language can help you study more effectively and process information more easily, especially during class lectures. Can you imagine taking a foreign language course and walking into class every day without having reviewed the terms used in the last class? Would you be able to follow the conversation for that day? What are the most popular study aids and techniques in foreign language classes?

- Note cards with definitions of terms on them
- Long lists of translated words
- Making a conscious effort to speak the language correctly
- Speaking translations out loud as you write sentences

These kinds of aids and techniques can be used in math class. In the course of this book, we will show you how you can use them in a way that will to maximize their benefit.

So, if math is a foreign language, why isn't more emphasis placed on teaching you how to speak it? Well, in regular foreign language classes, being able to speak the language is the end result; that's the goal. But in math, speaking the language is only a means to an end. The end result is to understand the concepts and develop the skills that the language expresses. For example, suppose an instructor writes the following statement on the board.

$$\text{The } \underline{\text{ordered pair}} \ (-2, 5) \text{ is not a } \underline{\text{solution}}$$
$$\text{to the } \underline{\text{system of equations}} \begin{cases} x + 2y = 8 \\ 3x - y = 6 \end{cases}$$

Professors hope that students understand the words underlined, but their primary concern is that students understand the *concept* being explained and that students can demonstrate the skill involved. Since concepts and skills are the primary concerns, they are the primary focus. The bottom line is that professors, particularly at the college level, do not have time to approach math as a foreign language course. You, on the other hand, have no choice but to approach it that way, at least as part of your studying.

Checking Out Vocabulary

Pick a section in your math book, preferably the one you are currently studying. Go through the section and write down all the boldfaced words in the space provided below. Then beside each word, write down a brief, "in your own words" definition. For this activity do only one section, not an entire chapter. A typical textbook section will have anywhere from 4 to 10 boldfaced or boxed words. I just want you to get a feel for how easy and beneficial this process is. Now, go for it!

Vocabulary for Section ....

....

....

....

_____

_____

_____

_____

_____

Hopefully, you get an idea of how effective this is. Taking the time to make a glossary of vocabulary terms is an essential aspect of your studying, regardless of whether or not your instructor requires you to do it for a grade. It will help you "talk the talk" in lectures, during homework, and on exams. In Chapter 6: Your Class Notebook, I will show you how to keep your vocabulary terms organized so they don't become clutter stuffed into your textbook.

Math Is a Skill-Based Subject

What do I mean when I say, "math is skill-based"? I mean that it must be _practiced_ repeatedly and not just seen to be mastered. Doing math proficiently is a lot like hitting a baseball. When your math professor is standing up at the board lecturing, he or she is showing you how to hit a baseball. You may watch and say, "Yes, I understand why she did that," or "I've got it!" But until you step into the batter's box and try it for yourself, your skill will not increase. Does this statement from one of my students sound familiar?

> _"When I'm in class and listening to a lecture, I understand everything. I get it. But when I get home and try the homework it's like I've forgotten everything and I don't know what I'm doing."_

I hear this all the time. What this student is missing is that you don't really "get it" when it's being presented in a lecture, even if you think you do. You only "get it" through repeated practice, just like hitting a baseball.

Understanding this skill-based idea will make math much easier, _if_ you act on it. Don't make the mistake of thinking you understand something until you have done it _on your own_. Also, you should make an effort to study math every day. Instead of studying for four hours every Sunday, you should study for an hour every day and take Sundays off. It also means that you should study as soon after class as possible, while all the ideas are still on your mind. Many students have demanding schedules that make it difficult to commit to studying every day. But even if you can't stop yourself from having to study in four-hour blocks, try to look at your math every day even if it's just for 15 or 20 minutes. This will help keep the ideas turning in your mind so that when you sit down for your next study session the material is fresh and you're ready to get down to business.

College Course Structure

A big part of what makes college math different is. . . well. . . college. That is, many of the difficulties facing students who are entering a college level math class for the first time are a result of the huge difference between a high school class and a college class. With the help of my students, I have compiled a list of the most important differences.

Differences Between a High School Class and a College Course	
HIGH SCHOOL	COLLEGE
Attendance is required.	Attendance is optional and may not even be taken.
Teachers monitor progress and performance closely.	Students receive grades, but are often not informed by the professor when they are in trouble.
Beginning algebra is covered over the course of 10 months for 6 or 7 hours a week.	Beginning algebra is covered over the course of 4 months for 4 or 5 hours a week.
The students have contact with the instructor every day.	The students often meet with the instructor only twice a week.
Teachers cover all material for tests in class through lectures and/or activities.	Students are responsible for information whether it is covered in class or not.
Teachers give frequent tests and often allow make-ups or retests if the grades are poor.	There are often only 2 or 3 tests a semester and no make-ups or retests are allowed.
Grades are often based heavily on level of participation and effort.	Grades are usually based exclusively on quality of work and demonstration of college level thinking.
A grade of "D" is interpreted as "barely passing" and the student is moved to the next-level course.	A grade of "D" is interpreted as "barely *not* passing" and the student is required to retake the class.
Teachers often offer extra credit for struggling students.	Professors almost never offer extra credit.

Because there is no attendance taken in many college classes, there is a temptation to skip class. Going at such a fast pace makes the material seem more confusing as it piles up day after day. You may or may not be surprised at the number of students who approach professors after a semester of poor grades, assuming that the instructor will do something to help them out just because that's what their high school teachers did.

Understandably, the differences between high school and college make quite a challenge for the student who never really "got it" in high school math class. The only way to successfully make the transition is by effectively managing your time, drawing on self-discipline, and being serious about your education. Time management is a skill that deserves major attention, so we will look at it in detail in Chapter 5.

Attitudes Toward Math

Students are taught certain stereotypes from a young age that influence their attitudes about math as a subject. Here are some examples of common "math myths":

1. Math is a subject that is *supposed* to be struggled through.
2. Math is boring.
3. To succeed in math you have to have a "mathematical mind."
4. Men are better than women at math.
5. It's okay to fail in math.
6. Math is not a practically useful subject and is never used.

These ideas are complete myths and have no foundation in reality. Here is the truth:

1. Math is well within reach of the vast majority of college students.
2. Math has a dynamic range of applications from psychology to anthropology to politics. In fact, no other subject even comes close in offering a more extensive set of tools for application and intellectual development. If math is boring, then education is boring.
3. The fact that you are *homo sapiens* (human) means you were born with a "mathematical mind."
4. Men are no better than women in math. If anything, women's tendency to be organized gives them an advantage.
5. It is only okay to fail in math if you think it is.
6. Our President *did not* stand up at the last State of the Union Address and say, "Our nation will fall behind technologically and economically if our college students don't learn more history. . . or biology. . . or

P.E." He said they need math. If the education system could only teach you one thing, it would be English. If it could only teach you two things, they would be English and math.

Many students understand that negative attitudes are false beliefs and that they need to avoid them but still the thoughts disrupt their study habits. In Chapter 4: Math Anxiety, we will focus on the anxiety caused by these attitudes and how to deal with them.

> **If you want to make small, incremental improvements in the way a person or organization performs, work on their behavior. If you want to make huge, quantum leaps in improving the way a person or organization performs, work on their attitudes.**
>
> —STEVEN COVEY, *The Seven Habits of Highly Effective People*

A Student's Math Background

Most people are aware of the fact that math topics build on each other. If you don't understand Chapter 1, you're probably not going to understand Chapter 2. But what students often do not realize is that this is also true for entire courses that follow each other. If you don't understand Prealgebra, you most likely will not succeed in Beginning Algebra. This makes the math class you begin with in college very important. Many students are in a rush to get their math requirement out of the way so they tend to take the highest level course they can. This is most often the wrong way to go about it, particularly if you keep in mind the demanding structure of college courses, as we discussed earlier. If there is any doubt in your mind about the level of math you should start at, then you should probably start with the lowest level course you are considering. Most colleges have a system for placing students into an appropriate math class. Suppose, for example, the placement test tells you that either Pre- or Beginning Algebra would be right for you. That means there are skills in Prealgebra you still need to learn. I guarantee that you will need those skills in Beginning Algebra. So if you take the lower level, what's the worst that could happen? You get to review the material, get a good grade, and now you're all ready for Beginning Algebra! Some students, however, are under time constraints and need to get through the course sequence as quickly as possible. In that case, you could choose Beginning Algebra, but remember that it will require more studying on your part to pick up any Prealgebra skills you may be missing.

This advice is not an idealistic notion I conjured up; it is the result of the experiences of dozens of students. This is what I have seen: when students take a higher level math course just to save time (or to get away with taking less math), they end up struggling in misery, failing the course, or dropping it. And the ones who go for the lower level class get a good grade and are ready to succeed in their next class. This has been my experience without exception.

So which is more fun?

- Struggling or even failing a hard class and having to take it again the next semester

- Passing a reasonable class and, at the same time, getting totally prepared for the next class

What if you find yourself in the middle of a class for which you know you are inadequately prepared? It can happen as a consequence of your own over-ambition, or perhaps the poor judgment of a counselor, or just bad luck; but it happens. Here is a list of suggestions for dealing with the situation:

- **Tell your instructor.** Often, an instructor can help you get into the prerequisite course even if it's past the deadline for adding the class.

- **Get a tutor.** Most colleges have tutor centers for math. A private tutor will cost about $15 to $20 per hour. You should never choose a tutor who is not referred to you by another student or who does not have student references, unless you're really desperate or feeling lucky.

- **Get a supplementary textbook from the previous course to help fill in the gaps.** It doesn't have to be the one the school uses, either. Brand new bookstore textbooks cost an insane amount of money. But you can find older math textbooks at thrift stores and used bookstores for $2. This is great for helping refresh you on things like how to work with fractions, solve equations, or add and subtract with negatives.

- **Get busy!**

"But I Had All This in High School"

I intentionally did not put this in the section on attitude because I do not think this point necessarily reflects a bad attitude. It is actually a pretty good point. Most (though probably not all) of the material in developmental college math courses is identical to material covered in various levels of high school. But the question is not "Did you *have* this in high school?" The question is "Did you *learn* this in high school?" And even if you did learn it in high school, are you ready to incorporate it as prerequisite material in college level courses like chemistry, statistics, biology, and business? College placement exams are designed to answer this question for you. The placement exam asks you math questions you need to be able to answer in college courses and places you a math class according to your performance. Also keep in mind, as we mentioned earlier, that in college you will be learning math at a much faster pace than in high school. When you get to the higher level courses you will be glad that you got to start off with college math at a level you were familiar with.

Make sure that you don't use the "been there, done that" mentality as an excuse to establish poor study habits in college. Instead, use these courses as a springboard to achievement in the higher level courses. Establish good habits *now* and watch yourself reap the benefits throughout college.

Quality of Instruction

Ultimately, the responsibility for passing or failing your math class lies squarely on your own shoulders (where it belongs). Still, a good instructor can make a world of difference. In fact, many students believe there is nothing more important than a good instructor. If you are such a student, please consider a few words of advice.

Finding a good instructor is not something you can leave to chance. You should start by asking your friends which instructors they like or go to a web site like *www.ratemyprofessor.com,* where students rate professors from colleges all across the nation. If several of your friends recommend the same professor, then there is a good chance that he or she is popular and you may have a difficult time getting into the class. This means you will have to be proactive during the registration process. You may even have to work your other classes around that professor's class time.

Another idea that is well worth the effort is to go and meet a professor before you take his or her class. Some students feel this may be awkward or inappropriate. The fact is, most professors would not only find such a gesture flattering, but would be eager to have you in their class given your level of motivation. Go by the professor's office and ask to see the course syllabus. Ask how he or she runs a class. The voice in the professor's head will sound something like this: "Wow! This student is really motivated. I sure do hope she takes my class!" Personally, when a student who is this motivated comes to see me, I tell the student I will guarantee him or her a spot in the class. Be confident! What you are doing is a good thing. If you go to see the professor and he or she acts annoyed by your visit, then run as far as you can in the opposite direction and never, ever take that professor's math class. It's that simple.

If you find yourself in a class with an instructor you don't "click" with, there are certain things you can do to help make the class go more smoothly. For one thing, hang on to your syllabus. It has all the instructor's class policies: from grades to attendance. That way you know exactly what is expected of you.

The syllabus will also probably contain a course schedule. This will help you stay on top of the material. Another thing you can do is use the other resources available to you. Most school libraries have supplementary video lectures that go along with the course. Find out where the tutor center is. Finally, the skills discussed in this book are designed to help you become an independent learner. These ideas will help you succeed in everything from note taking to test taking, no matter who your instructor is.

Finding Professor Right

In the space provided below, list the three qualities you think are most important for an instructor to have and why. Really think about it.

Quality 1

_____

_____

_____

Quality 2

_____

_____

_____

Quality 3

_____

_____

_____

TAKING THE ADVANTAGE

As we mentioned at the beginning of this chapter, making yourself aware of what makes college math different can turn the distinctions into an advantage. Approach the course with these ideas in mind. I assure you that the rest of this book was written to address these issues head on.

3 Learning Styles

You have probably heard it said that different people learn math in different ways. This is true not only of math, but of everything. There are three primary learning styles: visual, auditory, and kinesthetic. Visual learners learn best through sight, auditory learners learn best through hearing, and kinesthetic learners learn best through being active.

Most college classes are taught by lecture with notes outlined on a blackboard. These classes are great for visual learners. If you are not a visual learner, it can be difficult to absorb the information as it is presented. You cannot expect instructors to alter their teaching styles to facilitate your learning style. But you can determine your preferred learning style(s) and adapt your study habits in and out of class to accommodate that style and turn it into a strength. The following survey will help identify which learning style is best for you. Then I will make recommendations about how you can alter your approach to studying to facilitate your learning style. Many of these suggestions came from students.

Keep in mind that it is common for students to be strong in more than just one of the learning styles. In fact, they may all appeal to you. You should, however, make an effort to identify which of the three styles (visual, auditory, or kinesthetic) is your strongest.

Visual Learning

Consider the following questions carefully and answer "yes" or "no" to each.

_____ If you need to remember something like a formula or concept, do you have to rewrite it to remember it?

_____ Do you find the boxes, graphs, and diagrams in your textbooks particularly useful?

_____ Is it difficult for you to understand what your instructor is saying if it isn't written on the board?

_____ When you start a math problem, do you imagine yourself doing the steps in your mind before you actually start the problem?

_____ When you study, do you rely heavily on the notes you take in math class?

_____ Do you try to visualize your notes in your head when you're taking a test?

_____ Do you prefer written instructions to oral instructions?

If you answered "yes" to a lot of these questions, then you are a visual learner. Visual learners are the most common type. Here are some suggestions for how visual learners should approach studying.

- Notes are very important for you. You need to make sure that they are organized and thorough. Pay close attention to the material in Chapter 8: Class Time and Note Taking.

- Professors and textbooks often suggest that people draw boxes, pictures, or diagrams in order to help them solve word problems. This advice will be particularly useful for you.

- Note cards are a good study tool for helping you remember concepts and formulas. We will discuss how to make them effective in Chapter 9: Retention and General Study Strategies.

- The process of rewriting important information will be very helpful for you. When you review your class notes try to write as much as possible.

- Modern textbooks make use of effective design and sharp illustrations. You may even want to consider getting a second textbook to reference. You can pick one up for $2 at your local thrift store or used bookstore.

- It will help you tremendously if you write in your textbook. Underline things, write in the margins, and use a highlighter. If you do this, you may not be able to sell the book back to the bookstore, but you won't have to take the class over, either.

XERCISE

Auditory Learning

Consider the following questions carefully and answer "yes" or "no" to each.

_____ Do you prefer listening in class to taking notes?

_____ Do you have difficulty following written solutions on the blackboard unless the teacher also verbally explains all the steps?

_____ Do you remember more of what you hear than what you see?

_____ Do you find that your learning is most productive when you are discussing problems with an instructor or classmate?

_____ Would you rather have someone explain a problem than read about it in your textbook?

_____ Do you ever find yourself saying numbers and/or steps out loud when you are doing a math problem?

_____ Do you prefer oral instructions to written instructions?

If you answered "yes" to a lot of these questions, then you are an auditory learner. Auditory learners are not quite as common as visual learners, but they are still common. Here are some suggestions for how auditory learners should approach studying.

- Notes are a little bit trickier for you. Having class notes is important for everyone, but you want to take advantage of your auditory learning style. One way to really facilitate this is to bring a tape recorder to class. Ask your professor's permission to record lectures (99 times out of 100 the answer will be "yes") and listen to them again later as you look over your notes.

- Consider adopting a shorthand note-taking system so you don't spend too much time writing. There is a section in Chapter 8: Class Time and Note Taking called The Less You Write the More You Listen, which will help you develop this shorthand.

- Sit near the front of the classroom so you can clearly hear your teacher without distractions.

- It is a very good idea for you to ask questions in class because when professors answer questions they tend to talk more than when they write on the board. Ask for clarification of steps when you get lost.

- Repeat things out loud to yourself when you study: formulas, concepts, how to solve certain problems, instructions for homework. Read your textbook out loud. Do everything out loud! When you study, make sure to do it in a place where it is okay for you to speak out loud without disturbing anyone.

- If you struggle with being able to write everything down that the professor puts on the board, find a classmate who takes good notes and ask if you can photocopy his or her notes after class. Just remember: this is okay every now and then, but do not rely on photocopying other people's notes too often or you will get lazy.

- A study group will be good for you because they are heavy on mathematical conversation. There is a section on study groups in Chapter 9: Retention and General Study Strategies.

- If you a strong auditory learner and you think you have a hard time remembering formulas and concepts, record those things on a cassette or some audio device and listen to it a few times. You will be amazed at the results!

Kinesthetic/Tactile Learning

Consider the following questions carefully and answer "yes" or "no" to each.

Looking back on your education, have you learned math best when there were hands-on activities involved?

Do you ever pace or change position a lot when you are doing your math homework?

Do you have a hard time verbally explaining how to do a problem even if you're sure you know how to do it?

Do you find that math ideas don't "click" for you until you do the problems for yourself?

If you are having a good study session, does it involve a lot of small breaks?

Do you enjoy figuring out math puzzles?

Do you connect best with math ideas when they are in the context of real-life experiences?

Do you have a tendency to doodle during math lectures?

If you answered "yes" to a lot of these questions, then you are a kinesthetic learner. Kinesthetic learners are the least common of the three types of learners, but that doesn't mean they are uncommon. Here are some suggestions for how kinesthetic learners should approach studying.

- You need to get creative and dynamic. You learn best when you are active, so you want to find ways to make your learning active. One thing this means is that you will be more successful doing problems than reviewing how to do them. You still need to take good notes and review them, but focus on doing problem after problem after problem.

- Your class notes will be extremely beneficial if you take the time to rewrite them after class.

- Have you ever heard of a stress ball? It's one of those soft, palm-sized balls that you squeeze. I'm not saying you are stressed out, but if you take one to class it may help when you're feeling restless or antsy.

- Any time you are shown how to do a problem, you should follow up by trying one on your own as soon as you possibly can. This is a good idea for all students, but it is essential for kinesthetic learners.

- Move! When you study, stand up or even pace while you think about a problem. Move around. Take one-minute breaks to do push-ups or sit-ups.

- In Chapter 9: Retention and General Study Strategies, we will look at a study technique called learning maps that is great for kinesthetic learners.

- Be creative! Go outside and trace a formula in the sand or physically walk out the formula in an open space. Make a cheat sheet (Chapter 9) and tape it to the treadmill so you can stare at it while you run. Have the cheat sheet laminated and tape it to your surfboard so you can stare at it while you wait for a wave. Make learning maps

(Chapter 9) on big sheets of construction paper with markers. Anything that gets your body in motion will also get your mind in motion.

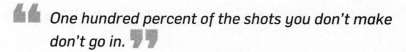 *One hundred percent of the shots you don't make don't go in.*

WAYNE GRETZKY, *Canadian hockey player*

 XERCISE

Techniques from the Learning Styles Survey

Based on the previous exercises, what is your primary learning style?

What is your secondary learning style?

What is your weakest learning style?

Look at the suggestions made for studying in your primary and perhaps secondary learning style and pick the five that you think would be most helpful for you. Then explain why you think that suggestion is a good idea for you.

1.

2.

3. _____

_____

_____

_____

4. _____

_____

_____

_____

5. _____

_____

_____

_____

4 Math Anxiety

TRUE STORY

When I was eight years old, one night I was having a nightmare about spiders and when I woke up there was a big, hairy spider sitting on my chest staring right at me. It was traumatic. The same month I accidentally hit a large nest of yellow jackets (bees) with my toy light saber and got stung 29 times (even in my mouth and on my eyelid!). Ever since then I have had a very intense fear of insects. Yes, when I see a roach, I jump on the nearest chair and scream for my wife. I'm sharing this because it is a good example of a situation where an experience early in life created an irrational fear. Now hopefully you are saying, "You're silly. Get over yourself," but this is most often the case with people who experience math anxiety; an experience early in life establishes an irrational fear. And in this chapter we'll discuss how you can "get over yourself" in regards to any anxious feelings you may have about math. Lucky for me, I didn't have to pick up a bug to graduate from college.

If you experience math anxiety, you are not alone. Research suggests that as many as 30% of college students struggle with math anxiety. And math anxiety ranks in the top concerns for college students. A good place to start is to try to pinpoint where you may have gotten this problem. To do this, I would like for you to write about your experiences in math class as far back as you can remember.

 XERCISE

Your Math Timeline and Biography

Timeline: Create a timeline of math classes and math-related events up to the present time, as far back as you can remember. In particular, track the classes you have taken and what your grades were for those classes.

Biography: In the space provided below, write the history of your experience with math. You don't have to re-list the classes you put on the timeline; this is an opportunity to reflect on how those classes went and why you think you got the grades you did. Also, if appropriate, discuss where things started to "go wrong" for you in math and any bad experiences you had. Discuss specific topics (such as fractions, variables, or equations) that you have had a difficult time with or have enjoyed. I have included detailed instructions and plenty of space because this is an important opportunity for reflection and we want plenty of it. Now, go for it!

SYMPTOMS OF MATH ANXIETY

Here are some typical symptoms of math anxiety. They may not be the only symptoms, but they are the most common, and the biggest threat to a student's success in a math course.

Avoidance

People who struggle with math anxiety often find themselves avoiding their responsibilities for math class, even if they are very disciplined otherwise.

- skipping class
- procrastinating doing homework
- not reading the textbook
- putting off math classes until the last possible semester
- choosing a major based on avoiding math
- cramming for exams

Avoidance and poor performance on math tests are not just symptoms of math anxiety, but also causes; they create more anxiety for the future. As we mentioned earlier, this "snowball" effect only makes the problem worse.

Poor Performance on Exams

People with math anxiety often find themselves "freezing up" on math tests, even though they have done everything in their power to prepare adequately. This experience can be very frustrating, and can have a "snowball" effect. That is, freezing on one test makes freezing on the next test even more likely. All of this occurs even though people with math anxiety typically do well in their other courses.

"I DO WELL IN OTHER CLASSES, SO WHY DON'T I DO WELL IN MATH?"

I get this question a lot. One reason has to do with the distinctions between math and other subjects that we discussed in Chapter 2: What Makes Math Different. Math anxiety is another big reason. People develop math anxiety for a number of reasons:

1. An embarrassing or unpleasant experience in an early class can trigger the fear that you are not good at math, or that math is a hard subject that can only be mastered by "smart" students. This is particularly dangerous for adults who had these experiences at a young age.

2. Society definitely perpetuates views about math that suggest that it is harder than it actually is. So once the seed of math anxiety is planted, society does its best to water it.

3. Since math is seen as more difficult, many people consider it to be a yardstick for intelligence. This causes pressure, either internally or from parents, to excel in math.

4. In Chapter 2 we listed several myths about math. Take a minute to review page 14. Research has shown that none of these myths are true, but many people still believe them. Math anxiety is a natural result that will follow from any of these beliefs.

THE CURE FOR MATH ANXIETY

Some amount of anxiety is helpful. It keeps you "on your toes." However, if your anxiety is so great that it is interfering with your ability to do well in class, there are steps you can take:

1. **Do math every day.** I cannot say enough about what this will do for you. Do math every day even if it's only for 15 to 30 minutes. Avoid stacking all of your studying into one or two days out of the week; if you study this way you simply forget too much in between sessions. Schedule frequent, reasonable study times and stick to them!

2. **Study smart.** Use the techniques in this book to make sure that you are taking a dynamic approach to your learning. The more you try different things, the more likely you are to find out what helps you most with memorization and understanding.

3. **Do NOT skip class** (unless it is absolutely unavoidable). If you have to miss class make sure you are following the syllabus to keep up.

4. **Be organized.** You need to keep a well-organized notebook for math class. Being organized will do a lot to stop you from feeling overwhelmed. I will help you develop an organizational system in Chapter 6: Your Class Notebook.

5. **Practice quizzing yourself.** In Chapter 10: Test Taking there is an entire section on how you can quiz yourself. Get into the habit of doing it. It is the best way to build your confidence. When you quiz yourself, it is most important that you don't "cheat" by looking at the book before you have your final answer.

6. **Replace negative self-talk with positive self-talk.** Be mindful of what you say to yourself in your head. Establish mantras—phrases you say repeatedly until you believe them! This may seem awkward at first, but be persistent. We will explore this in more detail in the next section of this chapter.

7. **Use your resources**. Go see your professor and let him or her know your concerns. Be honest. I promise that professors are well aware that math anxiety is a common problem. Find a study group. Seek out a tutor in your school's math tutor center.

The result of applying these steps consistently will be a marked increase in your confidence as a math student. This increase in confidence has its own snowball effect; the more you improve the more you'll want to do, which will make you even more successful and confident.

THE PHYSIOLOGY OF ATTITUDE

Math anxiety can cause negative thoughts to dominate your mind when you think about math class and your assignments. These negative thoughts lead to huge problems that go way beyond the idealistic notion that you should always have a positive outlook. That is, there is more to negative self-talk than just negative self-talk. First of all, negative thoughts create anxiety, and anxiety is a state of mind that releases chemicals in the brain that literally make it difficult for you to form connections between the parts of the brain that enable you to understand and perform. Negative self-talk can actually create chemical reactions in the brain that can make the attitude a "self-fulfilling prophecy." On top of all this, if you sit down to study and immediately go into a negative mantra about math, then every minute you spend complaining or being fearful is one less minute that you spend studying.

Often, a bad attitude about math is found in people who take their education seriously and are very disciplined in their studies. In spite of any bad experiences you've had in the past, decide right now that you have control over your success, that you will make every effort to succeed, and that you will be persistent in these efforts.

While it is a cliché, the fact is that you decide what your outlook on things will be. If you find yourself thinking a certain way, it is within your mental capacity to alter that course of thought to one that is more appropriate. You have the power to change "this stinks" to "this is a challenge I can meet." Change "math is boring and pointless" into "math represents opportunity." In fact, of all the factors that will determine your level of success, your attitude is the one over which you have the most control. The benefit you get from changing to a more positive outlook is obvious. Here is a list of positive affirmations you can use to help you get control of your attitude:

- I am responsible for the grade I make in this course.

- I will be patient and persistent in pursuing this goal.

- I am smart enough to pass math.

- I will pay the price for success.

Math Is Useless

I labeled this section "Math Is Useless" to get your attention. Now that I have it, completely banish the idea that "math is useless" from your mind forever. If our nation (or planet) could pick one subject for everyone to be really good at, it would be, without any doubt, mathematics. Math is the foundation for every science. Math is the ultimate in critical and analytical thinking. Math fires rockets into space, shows insurance companies what is a fair price, tells psychologists that therapy X works and that drug Y doesn't. It is true that nobody is going to hire you to factor a polynomial or solve a linear equation. But people are also not going to hire you to write essays on the Civil War, memorize every bone in the human arm, or draw sentence trees.

So why bother with an education at all? Well, when employers see a college degree they know the candidate has been exposed to and successfully assimilated a general, well-rounded body of knowledge. A college degree shows discipline, character, and sharp thinking skills. Apparently, that's pretty important. According to the U.S. Census Bureau, college graduates will earn $2.1 million more over the course of their life than high school graduates. Assuming that your degree takes four years, that's like getting paid $2,625 for every day you attend college! Work hard! Work smart!

> *An individual who is active in higher learning,*
> *Soon becomes an individual active in higher earning.*
>
> MICHAEL GULLIVER

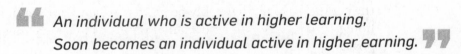

Yes, your textbook will give you word problems that involve two trains moving in opposite directions, which may seem irrelevant. These problems are there to sharpen your powers of reasoning. Keep this in mind. Your textbooks will also give you some very practical real-world problems that are only solvable using math. These are there to give you an appreciation for the power of math.

For many students who experience math anxiety, holding on to the idea that math is not a practical part of education serves as a justification for the behaviors that accompany math anxiety, particularly avoidance. It will help you with math anxiety to confront this belief and eradicate it utterly.

XERCISE

Self-Talk

Write down at least seven statements that you would like to use to replace any negative thoughts you have about math. You can use the information in the previous section as a guide, but make sure that the statements genuinely appeal to you.

1. ....

2. ....

3. ....

4. ....

5. ....

6. ....

7. ....

As you get deeper into study skills and begin to change your approach to the course, you will inevitably be more successful. Seeing the higher grades on homework, quizzes, and especially exams will make it easier to maintain your positive attitude.

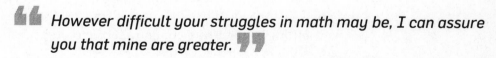

However difficult your struggles in math may be, I can assure you that mine are greater.

ALBERT EINSTEIN, *physicist*

5 Managing Your Time

> *A fruit never tasted so sweet as the one you walked furthest to pick.*

AUTHOR UNKNOWN

No book on how to succeed in college would be complete without a look at issues and strategies pertaining to time management. In fact, while this is not a manual about time management, for many students better time management is the single most important factor in turning around not just their grades in math class, but in college overall.

Many college students get lost in the social freedom that usually accompanies college and lack the knowledge of how to balance work and play. Other students are forced to pay for college or support a family while they attend college and, as a result, find themselves short on time to study. Whatever situation you're in, it will be of great value for you to reflect on where your time goes and think about some issues regarding managing your time. First, I would like to go ahead and bring to the table an idea that will be unsettling to some and common sense to others. But it is at the very heart of the time management issue.

THE COLLEGE RULE

The following principle is a generally accepted rule for college classes:

> *In order to realize your potential in a college level class*
> *you should study from two to four hours outside class*
> *for every hour you spend in class.*

I'll call this the College Rule. This is the reason why taking 12 credit hours makes you a "full-time" student at most institutions:

12 hours in class + 2·12 hours of studying
= 36 hours of commitment

And that's a minimum. This is the equivalent of a full-time job. Notice that the College Rule says, "to realize your potential," and not just "to pass." In some cases, realizing your potential may mean just managing to survive in a course. In other courses, realizing your potential may mean discovering something you're really good at and finding your dream career. Furthermore, the College Rule says that you should spend at least two hours studying. For classes that are more difficult, you may need to spend three or even four hours studying for every hour in class. Assuming that your math class meets five hours a week, that means you should be spending anywhere from 10 to 20 hours a week outside of class.

Course Time Commitment Evaluation

Think about all your classes and make a quick judgment about whether you think their difficulty level is Easy, Medium, or Hard. Try this quick survey to get an idea of how much time you should be spending studying every week.

_____ credit hours from easy classes × 2 = _____ hours

_____ credit hours from medium classes × 3 = _____ hours

_____ credit hours from hard classes × 4 = _____ hours

Total study hours per week = _____ hours

If you are not already studying this much, I urge you to resolve to do it for the next two weeks and see what the results are. Does this sound unrealistic? If so, then I would ask you why it sounds unrealistic.

Legitimate Concerns

In today's busy world we see many hard-working students who are balancing a number of responsibilities: school, part- or full-time jobs, children, military service, athletics, etc. Many students say they simply don't have enough time to dedicate to the College Rule. A student came to me last semester and complained that she wasn't successful in my math class. In the course of the conversation, I found out that she worked 20 hours a week, had a two-year-old daughter, and was taking 13 credit hours of college classes. I tactfully explained to her that if I were doing all that, I couldn't pass my own math course (and I write the tests!). If you have a particularly difficult time dealing with a certain subject (like math), then you may have to plan carefully so that you take it during a semester when your other academic obligations are minimal. If you can't avoid taking it during a time when you have too many other demands, then you could consider taking the course either online or as a self-paced course. Many colleges have set up such programs specifically to facilitate learning for students with major time constraints. Ask a counselor or math professor about these options.

Your Professor's Expectations

The time management issue presents a challenge to both student and professor. The student wants to do well but must operate under severe time constraints. The professor genuinely wants the student to succeed but can do very little (if anything) to help students commit to study time. Furthermore, every college professor I've ever met and talked to on the subject runs his or her class according to the College Rule. This is how professors set their expectations. Instructors are often blamed by students for being too demanding. There are three points that you, the student, should be aware of in regards to your professor's role in helping you succeed in a class if you do not have much time to dedicate to studying:

1. A professor is there to facilitate your learning in class. . . and that's all. Professors cannot give you a quick fix for the problem of "not enough time." The material covered in college courses is not determined by the professor, but by the school. So when I say, "They expect you to study for two to four hours outside of class for every hour in class," I don't mean your professor; I mean whichever college you chose to attend.

2. Only a very bad instructor would soften standards and make a class easier just to facilitate a student's lack of time to commit to the class.

3. While the two previous points sound uncompromising, keep in mind that most instructors have enough experience to advise you on whether you are more likely to sink or swim in your situation. And most instructors can tell you what you will need to put into a course to succeed after just one test. Instructors can also advise you on ways to maximize the effectiveness of the time you do have available to study. You only need to ask.

Student Motivation and Discipline

Given today's college environment I would not be being honest and thorough if I didn't address the issue of motivation and discipline in students. I'm talking about students who have the time to devote to the College Rule, but don't have the motivation. Professors call these students "slackers." If this is you (and you know who you are), rather than get preachy and resort to excessive moralizing, I'm going to plead with your better self. To you, the College Rule is unrealistic because it means sacrificing the majority of your leisure time. I understand this. I also understand that the prospect of changing this habit pattern and facing the College Rule is very challenging; it may even seem impossible. But, if you are not passing math for this reason, then your alternative is a few more semesters of having "fun" while you fail your classes followed by years of having an unfulfilling job because you didn't make it through college. Please *please* PLEASE, I'm begging you for your own sake. . . get into time management, face the music, and start giving the sacrifice that a college education demands. If you do, you may not ever love math (or even like it), but you will *pass* it. Then you will get your degree and the career you want and deserve. The bottom line is, you need to take an honest look at why you are in school, and decide if it is really what you want. If it is, then get busy. You can do it!

The College Rule and time management "pill" is a bitter one for many students. But, it is likely the most important pill in your prescription for success. A good number of students who feel that math is unmanageable are simply not aware of the demanding expectations in regards to their time management.

TIME MANAGEMENT STRATEGIES

Let's look at some techniques and principles that can really help you get a handle on how you spend your time and become more efficient with the time you have to study.

1. **Prioritize your activities.** Think about the things you do that take up your time: studying, work, television, friends, sports, shopping, etc. Then think about which of those are most important and begin your time management efforts with those priorities in mind.

2. **Set realistic goals.** Goal setting is an excellent activity to incorporate into time management. You could set a goal to have all your homework completed three days before an exam, then set aside times in your schedule to accomplish this goal. Be careful, however, that the goals you set are attainable. The most important thing about goal setting in the beginning is that you succeed with the goals you set. Otherwise, you will become frustrated and uninterested.

3. **Feeling listless? Make a list!** The "To Do" list is a wonderful tool for beginning to develop time management skills. You can make a "To Do" list daily or weekly or both. As with goal setting, you will be serving yourself best if you make it a point to make lists that you know you can complete for the most part. If you don't complete every item on the list, just transfer the unfinished items to the next day's list.

4. **Avoid "management by crisis."** "Management by crisis" is a term used to describe the way heavy procrastinators run their daily business. A "crisis" is a situation that requires immediate attention or there will be negative short-term consequences. Examples of this are waiting until the exact day that money is due for a traffic ticket to deal with it, or pulling a late night to get an important homework

assignment done. For people who manage by crisis, the day's activities are governed by what is necessary to avoid a crisis. Some people claim they work better under pressure. The problem with this is that while you may work better under pressure, you are also constantly stressing yourself out. The long-term effects of this can be bad for both your mind and your body.

5. **Just say "no."** A famous philosopher once wrote, "Things that matter most must never be at the mercy of things that matter least." Often friends make social requests that we are not genuinely interested in, but we say "yes" because we simply don't like to say "no." Of course, friends should be a priority. But spontaneous, unplanned social events should definitely take a backseat to being successful in college. Learning to say "no" politely and tactfully frees up time for "things that matter most."

6. **Don't forget to reward yourself.** Discipline is not the art of *avoiding* gratification. Discipline is the art of *delaying* gratification. As you set a goal, also set a reward for yourself when the goal is accomplished. This will keep you motivated. Rewarding yourself is an important part of sticking to a time management regiment. However, if you are diligent, you will eventually find that time management is its own reward.

7. **Use a day planner.** A day planner is an advanced type of "To Do" list. I thought it would be a good idea to include a sample of a weekly planner in this manual. It's on the next page. To use a daily planner, follow these guidelines:

   - On the Sunday before the week starts, fill in all the time slots when you know you will be busy with school, work, travel, etc. Don't forget to fill in any appointments or meetings that you know you will have.

   - Next, decide what times you would like to study and fill those in. Use the Study Hours Survey at the beginning of this chapter to help you decide how many hours you should try to set aside.

   - Plan other activities such as exercise and social events around the times already established.

   - Be flexible. That is, if something comes up that is unavoidable or genuinely irresistible, be willing to rearrange your schedule somewhat. However, be careful about what you are willing to call "irresistible" or you will end up procrastinating. A movie with friends is not "irresistible."

   - Stick to the schedule! At the end of the week you will look back and marvel at all the things you have accomplished.

If you like the sample planner shown here, make copies and use one every week. Or better yet, go out and buy a calendar day planner. Most people who invest time and effort into a day planner end up wondering how they ever got along without it. If you'd like to try using a planner but you don't like the one provided here, go online to *www.google.com* and do a search on "day scheduler" or "day planner." You will be find a dozen different kinds that you can print out and try.

Sample Day Planner							
Time	Monday	Tuesday	Wednesday	Thursday	Friday	Saturday	Sunday
7:00							
7:30							
8:00							
8:30							
9:00							
9:30							
10:00							
10:30							
11:00							
11:30							
12:00							
12:30							
1:00							
1:30							
2:00							
2:30							
3:00							
3:30							
4:00							
4:30							
5:00							
5:30							
6:00							
6:30							
7:00							
7:30							
8:00							
8:30							
9:00							
9:30							
10:00							
10:30							
11:00							

If you don't know where you are going, you might wind up someplace else.

YOGI BERRA, *Baseball Hall of Famer*

When to Schedule Math Study Sessions

As we have mentioned, math will be easier to learn if it is taken in small bits rather than big chunks. Ideally, you should schedule at least one hour every day to study math. Most people learn better and can think more clearly in the morning, but plan the time whenever you can. Again, you should make every effort to avoid studying in large blocks.

Setting Goals

Write down three goals for your math class. Make sure that the goals are attainable, but also make sure they will test your abilities. It may not be in your best interest to make one of your goals "Make an A on the next test." How about, "Pass the next test," or "Have my homework complete three days before the test," or "Study for an hour and a half every day."

Goal 1: ....

Goal 2: ....

Goal 3: ....

Evaluating Your Available Study Time

On the next page is simpler version of the day planner shown earlier. Use it to take a minute and write down all the obligations you know you have through the course of a week: work, classes, travel, etc.

Time	Monday	Tuesday	Wednesday	Thursday	Friday	Saturday	Sunday
7:00							
8:00							
9:00							
10:00							
11:00							
12:00							
1:00							
2:00							
3:00							
4:00							
5:00							
6:00							
7:00							
8:00							
9:00							
10:00							
11:00							

Add up the total number of hours these obligations represent, then add 56 hours to that to account for sleep. There are a total of 168 hours in a week.

Total Obligations: _____ hours + 56 hours = _____ hours

168 hours − _____ hours = _____ hours

This last number represents the total number of hours you have available to work out study time.

6 Your Class Notebook

Before we begin the discussion on how to organize your course materials, it will be good for you to reflect on your current organizational style.

 **Assessing Your Organizational Style**

Write a paragraph in complete sentences assessing your current level and/or style of organization. Be honest! Also, tell how these habits either help you or hinder you in your math class.

_____

_____

_____

_____

_____

_____

_____

_____

 We are what we repeatedly do. Excellence then, is not an act, but a habit.

ARISTOTLE, *Greek philosopher*

I cannot say enough about the importance of being organized in a math class. Being organized with your class materials, homework, and exams will turn math from an ocean of confusion to a discrete, ordered set of concepts that can be accessed for reference exactly when you need them. This chapter will address how you can set up a notebook that will keep you organized in a way that will make the course much more manageable and enjoyable.

This is a chapter in which some students may find information they do not feel the need to act on. Many students who struggle in math are already meticulously organized. If you are that type of student, read the information here to see if there is anything you can incorporate into your already effective style of organization. Also, if your instructor already imposes a notebook structure on you, then you should follow that format. If you are not an organized student, then I am telling you that you have no choice. Whether you choose the format I am laying out here or some other format, you must get your course materials organized.

THREE RINGS. . . NO CIRCUS

The organization system I am going to propose requires the use of a three-ring binder. Be careful. Make sure it is one that is going to be functional for the entire course. Students often buy the cheapest binder they can find and it becomes clogged with papers or difficult to close or insert new sheets into. In my students' experience, the thinner ones are actually easier to keep neat and are more practical. Do not use this binder for any other class besides math. Some students like to keep one gigantic three-ring binder for all their classes. If you are an extremely organized person this may work, but I advise against it. Trying to organize all your stuff into one large binder makes it difficult to keep the notebook functional. The format I am going to recommend is detailed enough that it needs its own binder. Read on and you'll see what I'm talking about.

The Structure of Your Math Notebook

Now that you have a three-ring binder, we can talk about how to organize it. The outline here can be modified to suit the needs of a specific student or instructor. Let's talk about what is important and should be organized into the notebook, and what can be discarded.

The first page should be a title page. This page is a good chance to store some important information that you may need to refer to during the semester. To help you accomplish this, I have created what I consider to be an appropriate title page; it appears at the end of this book. Notice that it contains information about your instructor, resources for assistance in your class, peer contacts, and a space to record your exam grades. This page has been perforated so you can remove it for use as the first page in your notebook.

The rest of the notebook should be divided into five sections. Each section should be separated from the rest by a labeled divider. This way you can navigate your notebook quickly and efficiently. Otherwise, you will waste time and become frustrated trying to find things. Here is a breakdown of each section.

1. **Handouts.** This section is where you will keep any handouts that are given in class. The first handout should be the course syllabus and class schedule. Most instructors give out at least a few handouts over the course of a semester. If you miss a class, be sure to get any handouts you missed from the instructor or make copies of a fellow student's handouts. In addition, if you find any resources that you think may be helpful to you in the course, such as study skills information off the Internet or an algebra summary sheet, you can put them in this section.

2. **Class notes and class work.** Include all your notes from class in chronological order. Each day's notes should be started on a new page with the date and section(s) covered on the top right-hand corner. We will discuss note-taking strategies in more detail in Chapter 8: Class Time and Note Taking. If your instructor incorporates group work or activities into class time, then those can be included in this section as well if you feel it will be beneficial.

3. **Homework.** These pages will hold the assigned homework. You should start every new section on a new sheet of paper. If your instructor collects homework assignments, simply take out the homework assignment to turn it in and put it back when it is handed back. Keeping your homework in such a neat manner will encourage you to take pride in its presentation.

   - Show every step.
   - Circle answers where appropriate.
   - Do not write in pen so you won't have to put huge marks through incorrect work.

   If you are the kind of student who uses additional practice problems beyond the assigned homework to help you study, that is awesome! But, for the sake of keeping things concise, you may want to store those somewhere other than in this notebook. The homework assignments given by your instructor usually represent everything you are expected to know. Often when students keep every single problem done, they wind up a messy notebook stuffed full of scrap paper. In the long term, only very neat, relevant homework should be kept in this section.

4. **Exams and quizzes.** In this section, put each exam or quiz in order as you take them and get them back. Correct each problem you missed on every exam. Your corrections should either be made on the test in a colored pen or follow immediately after the exam on notebook paper. I'm not suggesting that you retake the test, but that you have the correct steps and answers written down. You could get these from a classmate or from your instructor. The "Exams and Quizzes" section in your notebook is going to accomplish several things. Taking the time to three-hole punch the exam and look back over it will help you retain the information. Obviously, making the corrections will help with retention and understanding, particularly for the problems you missed. Finally, the exam section will serve as a wonderful tool for studying for the final exam (which will most likely be cumulative and based on these previous exams).

5. **Glossary.** In this section, you will create a list of definitions that come up in the class. The words you enter will have three sources:

    *   Instructor lectures: Most professors write vocabulary words on the board. Whenever they do this, you can either turn to the glossary and make a quick entry right then or, if you don't have time to do that, circle the word in your notes so you will remember to make it an entry in the glossary later when you review your notes.

    *   Textbook reading: In Chapter 7: Your Textbook and Homework, we will cover how to approach the textbook in a way that will make it easier to follow. One thing we will encourage you to do is add **boldfaced** words in the text to your glossary. Often, these words will appear in a list at the end of the chapter.

    *   Homework: As you do your homework, pay attention to key words that appear in the instructions or in finding solutions. If you have not done so already, you should add these words to the glossary.

When you add a word to your glossary, you should try your best to write the definition "in your own words." It will be easier for you to assimilate the definitions if they are worded in a way that you can relate to. At the same time, you need to make sure that your definitions are accurate. You may want to try putting in the textbook's "official" definition and then adding a few words that help you understand in your own way. The glossary section of your notebook should be reviewed often. As you make new entries, it will be natural to check out the ones you have already made. So the purpose of the glossary is served in large part just by your creating it. Remember that the goal is for you to get better at speaking the language of math.

General Remarks about Math Notebooks

One of the best things about the three-ring binder system is that everything is right at your fingertips. Students often ask instructors questions about the course that are answered in bold letters on the syllabus. This will not happen to you. You will never find yourself saying, "I know I have that somewhere." It's all right there. And there is a place for everything that could come up. Just find a convenient place on your campus where there is a three-hole punch, or buy one for yourself.

Speaking of a three-hole punch, make sure you do not get into the habit of putting loose-leaf pages in the pouches at the front of the notebook. Anything worth putting into the notebook is worth being punched and put into the appropriate section.

Having just one spiral notebook is simply not enough for a math class:

- Where do you put your syllabus and course schedule?
- Where do you put returned homework?
- Where do you put returned tests?
- What do you do if you want to let classmates borrow notes or you need to photocopy theirs?

Spiral notebooks are not versatile enough to organize all the information and material involved with being an excellent math student. Your mindset should be that your notebook is like a portfolio. It is very important that it not fall into disarray. Do not let the notebook get into a state where sheets are sporadically sticking out. You will learn much better from a notebook that you are proud of.

Bring your notebook to class with you every day. Show your notebook to your professor to get tips on effective studying for his or her class. The professor is likely to be so impressed that he or she will give away good information about tests.

I believe in this notebook system so much that I impose it on every algebra student I teach. The most common compliment I get from past students is that they now use a similar notebook system in all their other classes and that it makes a world of difference.

Exploring the Notebook Format

Referring to the format for the class notebook outlined in this section, list the five sections again here and explain why you think they would or would not make a difference in your ability to succeed in your math course. Don't be afraid to be honest.

Section 1:

Section 2:

Section 3:

Section 4:

Section 5:

"What If My Notebook Gets Too Full or Starts to Get Cluttered?"

It happens to the best of math students. In an effort to be the best they can be in math class, they accumulate so much material that their notebooks become bloated and unmanageable. Here are some tips to help keep you on track:

- If you take up a lot of space when you do homework and/or take class notes, consider removing the old assignment(s) and/or notes. Use a paper clamp to bind a chapter's worth of material together and store it somewhere you can find it easily later if needed. This will not be necessary for exams and the glossary since they don't take up much space.

- If the holes in some of your pages are tearing and causing the pages to stick out, make an assessment of whether the page(s) really need to be in the notebook or could be filed away. Or consider transferring the information to another page. Stay on top of these loose pages; nothing will make you lose interest in your notebook faster than ugly loose pages sticking out of it.

- Remember, your notebook is not a "catch all" storage space; keeping it clean and lean will make it the effective tool that it should be.

Be persistent in keeping your notebook neat and organized. It is very common for students to start out with a nice notebook and have it deteriorate over the semester. This is often a reflection of their performance in the course. Spend time with your notebook. Take pride in it. Love it and it will love you back.

7 Your Textbook and Homework

Let's be honest: few things in this world have confused and frustrated more people than math textbooks. Math textbooks have an unprecedented reputation for being the most confusing and boring pieces of published material around. As much as you may not wish to believe it, however, this reputation is undeserved. Not only that, your math textbook should be one of your best buddies. Think about all the wonderful things a math textbook contains:

- A definition of every word your instructor uses to confuse you
- An example problem that is almost exactly like the homework question you can't figure out
- The answers to half of your homework assignment in the back!
- A concise set of review problems at the end of each chapter to help you make sure you understand the chapter as a whole
- An exam at the end of each chapter that most times looks a lot like your instructor's real test

The problem with math textbooks is not that they are unapproachable but in how they are approached. Most students take an approach that is doomed from the start. They try to read their math book the same way they would read a novel: open up to the first page and just start reading. The problem is, of course, that your math textbook is not a novel and is therefore not likely to keep your interest. Have you ever been able to read more than a few pages of a boring novel? I can't. Reading to gather and assimilate information requires a totally different approach than reading for pleasure.

XERCISE

Your Current Textbook and Homework Approach

Make an honest assessment of how you currently use your textbook and approach your homework. Include as much detail as possible.

A good amount of research has been done on the most effective way to process textbook information. The technique is best summarized in a five-step approach.

HOW TO READ A MATH TEXTBOOK

These five steps should be followed for every section. An entire chapter is simply too large for these steps to be effective. You should, instead, approach each section individually in this manner.

Five Steps to Understanding a Section in Your Math Textbook

1. **Survey the material.** Begin the section by taking 5 to 10 minutes to glance through the material. For this step, you will not do much actual reading at all, but here are a few things you definitely should read:

   - Paragraph or section titles
   - Boldfaced words (although not necessarily their definitions)
   - Instructions for examples (but not the solutions)
   - Information that appears in boxes

   The idea behind this step is that you will begin to get an overall idea of what the material is about. You will know whether graphing is involved. You will know whether fractions are involved. Certain words will sound familiar, others will not. If you survey the material and everything you see looks totally confusing, that's okay. You have still done your job.

2. **Survey the homework.** Next, turn to the page that contains the section homework and see what kind of problems you will be doing in this section. This should only take about five minutes. In particular, read the instructions for the different problem types. If this seems overwhelming, you may want to try reading every fifth problem (1, 5, 10, 15, 20, . . .). This will give you a good overview of the whole homework set. Or, if your instructor has already given you homework for that section, you could read those problems. In these first two steps, you will be accomplishing several important things:

   - Looking at key words
   - Getting an idea of the big picture
   - Getting to know expectations

   Again, if you read through these problems and say to yourself, "Well, I just have no idea whatsoever how to do any of this stuff" that's okay! The fact that you have surveyed the material and its corresponding homework is going to help when you actually read the section.

3. **Read the section.** Now you're ready to actually read the material. But don't just read it. Your mind needs to be engaged with the material beyond the voice you hear in your head when you read. As you read it is important that you actively keep your mind focused on the topic. There are several ways to do this. Here are a few suggestions:

- Write boldfaced words in your glossary as you go. If you don't completely understand the term yet, leave some space below the entry so you can write in more later.

- Copy down the examples in the book. You don't have to write down every word of the instructions or every word of explanation in the solution. But if you at least write down all the math steps that are done it will help avoid that "Why did they do that?" feeling. Many times our eyes just move too fast to see math steps that should be simple. If you write down the math steps in the examples, it will be easier to see why they were performed. Think about it; it will also be like training wheels for the homework in that section.

- Write down anything that appears in a box, particularly formulas and equations. If it looks familiar and you know it will be important, you could write it on a note card or in a corner of the page where you know you will do homework. If you are a kinesthetic learner, take the time to doodle a formula into your notebook somewhere you know you'll see it again later (but be careful not to waste too much time with that).

- Read out loud to yourself. This is a great idea if you're an auditory learner. Make sure you are not disturbing anyone around you when you do this. But as long as you are not, go for it!

- Writing in your book is a great way to make friends with it. If you don't mind keeping your book after the course is over and not selling it back, there are several things you can do:

  □ Use a highlighter to make important information stand out.

  □ Put a question mark (in pencil) beside ideas or examples you have a hard time understanding.

  □ Circle key terms.

Using a highlighter is okay, but DO NOT use an ink pen to write in your book. Use a pencil so you can edit your comments later.

Remember: If you just sit there and read to yourself and that's all, you are not likely to be successful. Your brain needs to be stimulated in other ways to keep it active and interested.

4. **Survey the material one more time.** Now that you are done reading the material, take another five minutes or so to go back and review both the textbook and your handiwork. Look at the things you wrote down, and survey the section again. This will help with retaining the information.

5. **Homework time.** Now you are ready to go to the section homework and give it a shot. If your instructor has already given homework for this section, you should try it. If not, try every other odd problem (1, 5, 9, 13, . . .). That way you get a nice variety of problems without having to do so many of them. Doing every other odd problem takes a section's homework and cuts it down to about one-fourth the amount of the total problems. So, if a section has 80 problems in the homework set, doing every other odd problem would be 20 problems. You may need more practice than that, you may not.

Trying the Reading Technique

Attempt this technique for one section in your math book and give a summary of the results. Discuss any challenges, pitfalls, and/or breakthroughs. In particular, how did the homework for that section go? Do you think you will start to employ this technique on a regular basis? Why or why not?

"What If I Get Bogged Down in the Reading Step?"

If you have surveyed a section but find that you become confused when you try to read it, make sure you are trying some of the techniques involved with active reading. In particular, try writing in your book to fill in steps, write down questions you have, and make comments as you make connections. As we mentioned earlier, not being able to sell your book back to the bookstore is not nearly as painful as failing a math course.

DOING HOMEWORK

When they do homework, most students just sit down, open up the book to the homework page, and start on the first problem without looking over their notes or the textbook section at all. This is the mental equivalent of getting out of bed in the morning and going straight into a sprinting routine without doing a single stretch or warm-up. It simply won't be effective and is likely to leave you strained, sprained, or throwing up. It is important that you take the time to review what is happening in the course before you start homework. Your mind needs to be warmed up to the concepts and skills it will be asked to explore.

How to Do Homework

Here are seven steps that will make homework easier and more productive.

1. Begin by reading the corresponding section as outlined above. If you have already done this, take five minutes to review the section in the textbook before you start.

2. Review the professor's notes for that section.

3. Now begin the homework with the first problem. If a problem is too difficult, search the textbook and your notes for a similar example. Even if you don't completely understand why you are doing the steps you are doing, see if you can move through the process and get the right answer.

4. Record each step neatly. Use as much paper as you need. Space things out to allow for comments or questions about the problems.

5. NEVER, NEVER, NEVER do your math homework in pen. Ink is the number-one ingredient in the recipe for sloppy homework.

6. Use the "10 Minutes of Frustration" rule: If you become frustrated with a problem and you have genuinely exhausted every possible resource for trying to solve it, work for 10 minutes beyond that point and then, if you still don't get it, STOP! Put a question mark beside the problem and move on. Students come to me and say, "I

worked on this one problem for an hour and still couldn't get it," and I say, "Well, why in the world did you do that?!" While the effort is a valiant one, it is a poor study skill. Take the problem to someone who can help.

7. Always finish a homework session by doing a problem that you know you know how to do, even if it means taking a step back to a previous set of problems. This will end your session with your mind in a state of confidence.

"What If I Get Bogged Down Trying to Do Homework?"

It is important that we address this problem because sometimes students just feel as if they can't move forward at all. Remember that you are responsible for your success in the course, but stay positive! You are not a victim of math. You are an active learner on the straight and narrow path to academic excellence! If you have gotten most other topics, this one can be conquered too. Get up, take a 15-minute break, and then try again. Also, be proactive by remembering the resources you have available.

- Your professor may not be available at the moment but you should talk as soon as possible.

- Find out when your school's tutor center is open, and plan your study times so you can study there.

- Moving forward when you get bogged down in homework is a problem most easily solved by participating in a study group. Find classmates who want to succeed. Get their phone numbers and the times they usually study. They are likely to be experiencing the same difficulty. We will discuss study groups more in Chapter 9: Retention and General Study Strategies.

The Best You Can Be

Of the techniques mentioned in this chapter, which are the most appealing to you? Why? Are there any that you do not think would be helpful to you? Why not? Go back to the exercise at the beginning of the chapter and assess why you think your old approach was effective or ineffective.

....

....

....

....

All winning teams are goal-oriented. Teams like these win consistently because everyone connected with them concentrates on specific objectives. They go about their business with blinders on; nothing will distract them from achieving their aims.

LOU HOLTZ, *former Notre Dame football coach*

8

Class Time and Note Taking

NO PAIN, NO GAIN

Do you get bored in math class? If so, you are taking the wrong approach. Class time is often seen as a laborious exercise in the divulging of information. Instead, class time should be seen as an intense mental workout. This is why you're in school! How many times have you finished a long, hard workout and thought, "That sure was boring." Probably never, because during a workout you are so busy working hard that you don't have time to experience the feeling of boredom. Math class should be the same way. Perk up! Grab a cup of coffee if you have to. This is the time when you are going to be exposed to the new skills you must learn to succeed. This is the time when most of your learning will take place. Don't let it go to waste! Let's look at some points that will make your experience in math class much more enjoyable and profitable. But first, let's look at where you are right now.

 XERCISE

Your Class Time Habits

This exercise may take some personal observation on your part. Write a paragraph in the space provided below that describes your mindset during class. Are you bored? Where does your mind tend to wander? Are you anxious about the lecture or something else? Do you take notes? If so, how? Do you have any before-class rituals or after-class rituals?

If you're not really sure, take a couple of days to observe yourself, then come back to this exercise. If your habits have been affected by the material in this book, that's great! Just be honest.

BEFORE CLASS

Your success during class actually begins outside class. First, and most obviously, you need to have attempted all the skills discussed in the last class. This means doing homework. You can use this time to help you generate any questions that you have (we'll talk more about asking questions later). If your schedule is such that you haven't had time to attempt all the homework from the last section, at least go to the section and read through the assignment to see what types of problems are there.

The Warm-Up: 15 Minutes to Class Time

As a teacher, every day when I walk into class I see students talking about different things. The successful students are most often (though not always) talking about the homework or material. And nothing is quite so satisfying as walking in and seeing two or three students up at the board helping each other out. Other students talk about what's on TV or what they're doing this weekend, and they tend to be less successful. Of course, I am not suggesting that you only discuss math with the people in your math class. But I am saying that **successful students don't wait for class to start, they start it themselves.** Fifteen minutes before math class starts, you should be doing one of these things:

- Reviewing your notes from the last class. This is probably the best thing you can do to warm up.

- Surveying the section that will be covered that day. Even if you are clueless about the new material as you survey it, it will help during the lecture that you have exposed yourself to the list of topics and vocabulary words.

- Reviewing the material from the last section to prepare questions.

- Working with classmates on previous material, receiving or giving help.

- Doing one or two problems from a previous section that you know you can do. This can be a good, quick confidence booster and it really helps your brain warm up.

If your math class starts 10 minutes after another class ends, you should follow these suggestions before you go to school. It really is like warming up before exercise; it makes you feel better while you're doing it and it makes the exercise more productive. Resolve to try this for the next four class periods and see if it makes a difference.

Punctuality

When the time for class arrives, you should have paper and pencil out with your textbook and notebook readily available. Your eyes are forward and your attention on your professor. This is a matter of etiquette. Also, if you consistently take this approach it will likely be noticed by your professor. Nothing is more annoying for a professor than to have to get the class's attention because some people want to finish their conversation about how awesome their weekend was. It's not that we don't want you to have awesome weekends, but now it's time to get down to business.

Get a Good Seat

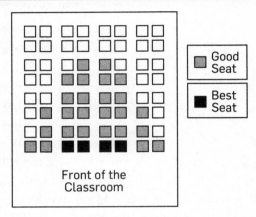

Sit in the front row, if at all possible. I have never had a student move to the front row and then complain about the view. Serious students tend to sit up front so they are not distracted by their classmates. This may mean getting to class early, but that's great because it will give you a chance to warm up. If you can't get a front-row seat, you should make sure you are more toward the center the farther back you have to go. Up front and toward the center is where you want to be. It will be easier to pay attention and not be distracted.

DURING CLASS

Okay, good; you have warmed up, you're in a good seat, and class starts. How do you stop your mind from drifting into never-never-land or spiraling into boredom? The best thing to do is become a good listener.

You can make yourself a better listener by making it a point to keep your eyes focused on only two things: your professor and your paper—not the clock, not the cute guy (or girl) four seats away, and not empty space. It can also help if you make an effort to become an *active listener*. By this I mean that as you listen you are thinking of potential questions and making notes on your notes that will guide you later when you look at them at home. This is obvious, but does not come naturally; it takes discipline and determination. Later on in this chapter, I will show you a system of shorthand symbols that can help you quickly mark your notes with prompts that will tell you where important parts are, and, if necessary, where you got lost during class.

Note Taking

Note taking is a topic that could (and does) have whole books written about it. A good bit of research has been done to determine what is most effective in this area. I am not going to push a note-taking system on you. I will suggest one system that I see as particularly useful and have you do an exercise to see how it works. But ultimately, how you take notes is up to you, and different things work for different people.

While I don't think any particular note-taking system is better than the rest, I will say that it is important that you do develop note taking as a skill. Also, you should always follow these guidelines, no matter what system you use:

- Begin each class with a new sheet of paper. Record the date at the top of the new page.

- Space your notes out. Leave gaps if you need to, especially in places where you're having a hard time understanding. You can fill these in later while you're studying. This is particularly helpful if you are not the neatest person.

- Copy all steps shown, even if you are clueless and even if you already know them. Do this especially for word problems.

- The less you write, the more you listen! Use abbreviations to save writing time. In a bit we will look at a useful abbreviation list.

- Don't stop taking notes, even if you get confused.

- Cues for important material come naturally to professors. That is, it is natural for a professor to become emphatic or to give clues about material that will definitely be on the exam. Listen carefully so you can mark these problems.

- Develop a glossary of key words to review periodically. This will work wonders for understanding math as a language.

- Don't be afraid to personalize your notes. Buy a highlighter or a box of colored pencils. Color-code the material to make it easier for you to follow.

Now let's look at one specific system for taking notes. You may decide this system is right for you; you may not. Regardless, being exposed to it will help you understand what it takes to be a good note taker.

The Column System

This system of note taking involves dividing your paper into vertical columns. Each of the columns is used for specific material. The benefit of this system is that it eliminates the tendency for student notes to be a chaotic swirl of information. Students often come to me with their notes out and say, "Didn't you do an example on this?" or "I think you told us what that meant but I don't see where," and the information will be sitting on the page right in front of them. The students aren't so much to blame for this; they have taken down everything I said. The problem is that there is so much "stuff" on the page that when they look at it it's just a dense mass of unsorted information. The column system allows you to pick one or two aspects of your notes and quarantine them so they aren't lost in the mix. Here are some examples of how you could set up such a system. Some of the formats involve two columns, and some involve three.

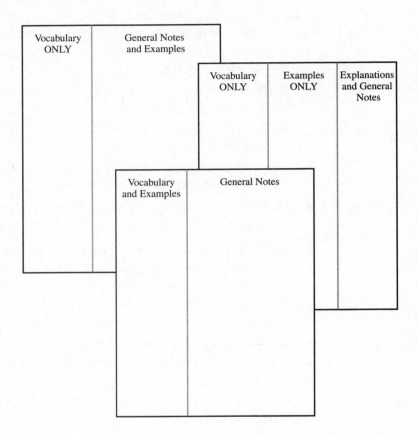

As you can see, there is a lot of flexibility here. If you have always had a hard time with vocabulary in math, then establish a thin column that contains only vocabulary. If you have a hard time processing examples, establish a column for only examples. Then, when you go back after class to reprocess those examples they will be isolated and you will have plenty of extra space in that column to make notes about how to do them. One important point that you may have already picked up on is that once you establish the contents of a column, you must NOT put anything else in that column. Otherwise, the whole point of the system will be defeated and your notes will go back to the swirl of chaos that they once were. Also, use a ruler to make the partitions neat.

Trying the Column System

The next time you are in math class, try using the column system of note taking and see if it helps. If you are unsure how to establish each column, then I suggest you try the two-column system of "Vocabulary and Examples" and "General Notes" as shown on the previous page. Try it for at least one class period. In the space below, discuss the effectiveness of the technique. Do you think you will start using this system on a regular basis? Explain why or why not.

_____

_____

_____

_____

_____

_____

Asking Questions

It is true that students don't ask questions because they don't want to look stupid in front of their classmates or their professor. It is also true that this fear is completely unfounded. However, I believe that while this fear may be unfounded, it is understandable, particularly for people who have math anxiety. So, I figured, rather than try to talk you out of being afraid of asking questions, I would show you how you can make sure that your questions are GOOD questions. But I should try at least once to talk you out of being afraid to ask questions, so here goes.

> **Don't be afraid to ask questions! Questions are crucial for learning. Your professor wants you to ask questions! Your professor will love you for it, and your classmates will think you're smart and brave for asking the question they were afraid to ask!**

Okay, now that that's out of the way, let's look at how you can frame questions so that they will be on point. Here are a few suggestions. Some are ways to begin your questions and others are full questions, but all of them will help you ensure that your questions are productive for you and the rest of the class.

- "Could you quickly go back to. . . ?"
- "Is that the only way to do that step?"
- "Could that step be done this way?"
- "Could we. . . ?"
- "What's the difference between. . . ?"
- "What if we. . . ?"
- "Why did you have to. . . ?"
- "For this type of problem, is that example about at the level of difficulty that would be on a test?" (Don't ask this one too much, but every now and then it's acceptable and professors are often pretty generous with their answers.)

Suppose you're doing a pretty good job of following along and suddenly you get lost.

BAD: "I'm totally clueless about what you're doing right now."

GOOD: "Okay, I followed up to this step, then you lost me."

GOOD: "I was following the lecture fine until you did this example. I got it down in my notes but could you mention anything else that may help me?"

IMPORTANT: If you ask a question and a professor tells you the question is too remedial or off base or there isn't enough time to answer it, do not take it personally. Again, professors love it when students ask questions. If they say they can't answer the question then they probably have a good reason. Don't let it stop you from asking questions in the future.

This concludes the section on ways to make your time during class more productive and enjoyable. Remember, successful students do not wait for their professor to begin class; they start it themselves by getting prepared and being active in the learning process. Arrive early, get a good seat, listen actively, take good notes, and ask questions. Class time is a workout; no pain, no gain. These things require discipline and focus; they're like mental kung-fu. Make your kung-fu the best!

AFTER CLASS

One of the hardest things about physical exercise is making yourself do a cool-down. But walking for five minutes or stretching after exercise is very important for maximizing the effectiveness of the exercise. It brings muscles out of their intense working state into a normal state conducive to relaxation. The same is true for math class, but in a slightly different way.

Have you ever heard a math student say, "When my professor did this example in class I totally understood what was going on, but when I got home

I was clueless"? You may have said this yourself. This happens frequently to students for a simple reason. The amount of time between the lecture and sitting down at home to do homework is more than enough time for the brain to forget all about what it learned. You can cut way back on this experience by using the first 15 minutes outside class to review your notes. The material will still be fresh in your mind and you can give your brain a nice opportunity to reinforce some connections.

Yes, I am saying that after two hours of sitting in class you should study for 15 more minutes, and I understand that this requires great discipline. Maybe you could review while you have lunch or a snack before you go to your next commitment. I guarantee that you will be saving yourself a lot of time when it comes time to actually attempt the homework. Reviewing notes after class will work wonders for your ability to retain information. When you sit down to do homework, instead of saying, "What was that again?" you'll think, "Oh yeah, I remember looking at that." Then you'll have more time for other things because you'll be done with your homework faster.

If you have another class immediately after your math class and can't commit to reviewing for 15 minutes, you should do it at the first possible opportunity, even if it's later in the day when you get home. The longer the gap between class and your first exposure to that material outside of class, the more likely you are to have forgotten what you learned.

USING NOTES OUTSIDE OF CLASS

Obviously, the better you are at taking notes in class, the more they will help you outside of class. One point we made about notes earlier is that whatever method you use for taking notes, you should leave yourself plenty of space. This is so you can go back later after class and rework the notes. When I say "rework" your notes, I don't mean "rewrite" them. I mean you should go through and do the following:

- Reread the notes.

- Add in extra steps and comments to the notes so that they are clearer.

- If you see something that you understand now but think you may forget later, clarify by adding a few notes to help your future self.

- If there is something you don't understand, put a question mark by it and try to think of a good question you could ask. Write the question down.

- You may want to circle vocabulary terms or examples that the instructor indicated as particularly important. But don't get "circle happy," or the circles will become meaningless.

The exercise of reworking your notes will accomplish several amazing things:

- Your ability to transfer things into solid, long-term memory will skyrocket!
- The material you are expected to learn in the course will become ordered and manageable.
- Your notes will go from being a confusing obligation to being your bread and butter.

THE LESS YOU WRITE, THE MORE YOU LISTEN

Let's face it, college professors can talk very fast. Any book on note taking will suggest that you develop a system of shorthand notations that will allow you to take down information faster. This is very helpful in math class. At the end of this book, on the back of the page provided as a title page to your notebook, there is a table you can use to begin a system of shorthand specifically for a math class. And there are some blank rows where you can fill in your own abbreviations. If you keep this sheet handy and make an effort to use the symbols, it will save you a lot of time in class. You may even want to print out an extra copy to keep in front of you during class.

The Best You Can Be

We began this chapter with an exercise about your current class time habits. Since then we have presented a lot of material. Now that you have been through this material, write a paragraph about what you think would be the most effective approach to class: before, during, and after. Feel free to mention things you have thought of that I may have forgotten to mention in this chapter.

9 Retention and General Study Strategies

HOW YOUR BRAIN REMEMBERS MATH

You brain is made up of about 100,000,000,000 (100 billion) neurons that work together to store and process information. The neurons are networked together by 100,000,000,000,000 (100 trillion) connectors called synapses. As you learn new information, clusters of neurons connect with other clusters of neurons by forming dendrites that further facilitate the transferring of information. When you first learn a skill, a weak dendrite connection is made between parts of your brain that store the relevant information. The more you practice a particular skill, the stronger the dendrite bond becomes. If you wait too long before you reinforce the skill, then the weak dendrite connection is lost.

Suppose you see your professor do the following example on the board. The example involves adding fractions.

Example: Add $\dfrac{4}{15} + \dfrac{5}{12}$

$\dfrac{4}{15} + \dfrac{5}{12}$

$= \dfrac{4}{3 \cdot 5} + \dfrac{5}{2 \cdot 2 \cdot 3}$ Factor the denominators

$= \dfrac{2 \cdot 2 \cdot 4}{2 \cdot 2 \cdot 3 \cdot 5} + \dfrac{5 \cdot 5}{5 \cdot 2 \cdot 2 \cdot 3}$ Multiply the numerator and denominator of each fraction to get the LCD

$= \dfrac{16}{60} + \dfrac{25}{60}$ Simplify. Notice that the fractions now have a common denominator

$= \dfrac{16 + 25}{60}$ Now that they have a common denominator, we can add the numerators

$= \dfrac{41}{60}$

So, suppose you see this example and you "get it." You understand all the steps. Congratulations! Your brain has made several dendrite connections. This problem involves multiplying, factoring, and adding. The clusters of neurons responsible for all these distinct tasks have been connected by dendrites to form a new cluster of neurons that knows how to add fractions. But, these new dendrite connections are weak. If you wait three days before trying to reinforce them and establish your fraction-adding cluster of neurons, the connection will have been lost and you'll say, "When I was in class, I got it, but when I got home I didn't understand." How can you facilitate the strengthening of your new dendrites? By practicing the skill as soon after the lesson as you possibly can, and practicing it repeatedly. This is why, back in Chapter 8: Class Time and Note Taking, I emphasized that you must review your notes as soon after class as possible, preferably immediately after class. And this is why we've been stressing the importance of doing math every day, or as often as possible.

This is what I mean when I say that the key to success is not in high intelligence, but in an intelligent approach. Many students think that they are just dumb and can't do it on their own and that's why they don't get it when they get home. Not so. Their brain made the connections; they just waited too long to have those connections reinforced.

This short lesson in brain biology resonates with many students I share it with. One student in particular framed his whole attitude around these facts. So he would come to me and say things like "The dendrites aren't forming on this problem" or "I had a strong connection in class but it's gotten weak." He also understood that he needed to practice, practice, practice in order to keep these connections solid. He would never let me do a problem for him; he always insisted that I talk him through the steps while he wrote them down. He went from being what he called "hopeless" to a solid "B" student in one semester.

Short- Versus Long-Term Memory

I include this section because it is an area of interest for many students, and I get asked a lot of questions about it. Many students will say that cramming for tests makes use of short-term memory. Actually, short-term memory only lasts for about 30 seconds. Any image, sound, or concept that enters your brain and is considered for real memorization passes through short-term memory. Practically, you use your short-term memory when you look at a number in the phone book, then sprint to the phone to call before you forget it. Any information that your brain remembers beyond 30 seconds has been transferred into long-term memory. Now it's just a matter of how solid the connections are. So, for example, when you cram for an exam you are making a huge number of very weak connections in your long-term memory. Obviously, they will not stick around for long, and soon you won't remember any of what you studied at all.

RETENTION TECHNIQUES

Retention is a major barrier for many college math students. That is, many students do well going through homework, then when the time comes to prepare for the test, they have forgotten a lot of what they have learned. The key to avoiding this is to find a good tool for memorization and employ it, not just when the time comes to begin exam preparation, but as you go through the chapter. We will give examples of several memorization techniques and do a few exercises to help you become more familiar with how they work.

Note Cards

Ah, note cards. Many students have had their performance completely changed by this technique. Buy a pack of note cards. Any size will do. On each one, put a vocabulary word, the title of a formula, or a problem, and on the other side put the definition, actual formula, or solution to the problem. Don't put more than one thing on each note card. This will serve two purposes:

1. It will turn an ocean of material into an organized stack of information in small bits.

2. It will make it easier for you to use the cards to quiz yourself.

Here is an example of the front and back of a good formula note card.

Effective Note Card

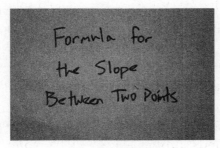 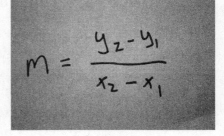

Front   Back

The title of the formula is on the front and the formula itself is on the back. The student will flip to this note card, then try to recite the formula on the back. This is GREAT for memory! Here is a good note card that has an example on it.

Effective Note Card

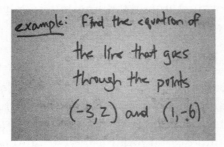

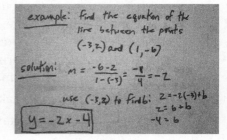

Front   Back

The question is on the front and the solution is on the back. The student will flip to this note card and try to do the problem. Once she has the solution she will turn the card over to check herself. This is a great way to practice word problems. Now let's look at an example of a note card that will not be as effective.

Ineffective Note Card

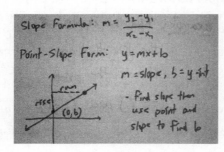

Front   Back

On the front we have two formulas, a graph to illustrate a concept, and a note about how to do a problem. On the back we have an example that makes use of the information on the other side. There are two problems with this note card:

1. The note card is not in a format that allows the student to quiz herself on a problem.
2. There is so much information on the card that the student may not get any more out of it than she would just reviewing her notes.

But if you only put one tiny piece of information on each card, won't that take a lot of cards?! Yes, it will. But don't worry, they're really cheap. Students who use note cards usually work with a hefty stack as big as two inches thick. Now that's a lot of note cards! And it works like a charm.

Note cards are effective for several reasons. The process of creating the cards makes use of all learning styles:

- You're writing the information down, so it's **visual**.

- You're active, so it's **kinesthetic**.

- If you want you can say the information out loud as you write it down, so it's **auditory**.

Also, the process of sifting through the material and deciding what is important enough to get its own note card is a powerful time of review. And that's just making the cards! Once you have them made, not only can you quiz yourself, but you can quiz yourself any time or place you see fit:

- while you do laundry

- waiting for the bus

- waiting for an appointment

- in the bathroom

- while you walk across campus

- NOT while you are driving

You don't need book, pencil, and paper to study. You just need your note cards and a rubber band to keep them together. This is another thing that makes note cards a highly kinesthetic activity.

Take your note cards to your professor and ask him or her to flip through them to see if the material is good. If you feel like you have so many cards that this will annoy your professor, then take out the ones that are obviously good. Maybe show all the note cards for word problems or the formulas you think you need to memorize. If you do this, be ready for praise and quite possibly a major exam hint.

Making Note Cards

You owe it to yourself to try note cards. I really want for you to give this a fair shot, so I'm not asking you to do a whole chapter, just a section. Choose a section from the chapter you are currently working on in your math class and make at least 15 to 20 note cards. Make sure to include a variety of vocabulary, formulas, examples, and word problems so that you get a representation of the whole section. After you have made the note cards, use the space below to write an honest appraisal of the experience. Do you think it would be helpful enough to you personally to start doing it for every section? Why or why not?

_____

_____

_____

_____

_____

_____

_____

_____

_____

_____

_____

Note cards are a very effective memory technique. They are, however, not without pitfalls. When you make note cards, make sure they are dynamic and that you didn't miss anything important. Also, make sure that once you have made them you don't completely abandon your textbook and notebook. There is still plenty of work to be done there.

Learning Maps

Making a learning map is a technique that is fun and appeals greatly to visual learners and particularly kinesthetic learners. A learning map is a summary of information put into diagram form, somewhat like a flowchart. You should do a learning map for each section in the chapter instead of trying to do one big learning map for the whole chapter. You begin by putting the section title in the center of a page and circling it. Then you connect the objectives in that section to the circle and branch details out from there. In this case, a picture is worth at least a thousand words. Here is an example of an effective learning map.

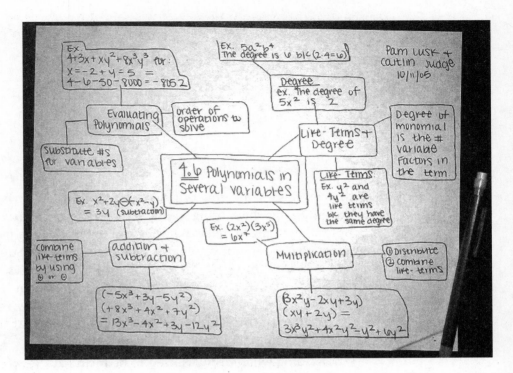

As mentioned, the main title is in the center and the topics for the section surround the center. Then specific aspects and examples of each individual topic are connected. Just be careful that your map doesn't get so cluttered that it becomes hard to follow.

It may seem like this is a lot to go through to get a summary of a section, but there are several things about making and studying learning maps that make them a very effective memorization tool. Learning maps are another highly kinesthetic activity. This process works wonders for transferring concepts, formulas, and problem-solving techniques into solid, long-term memory.

Learning Maps

As you did with note cards, pick a section from the chapter you are currently working on and create a learning map for it. Use the objectives at the beginning of the textbook section as a guide for the topics to include. Include vocabulary, concepts, formulas, and examples. In the space below, discuss how it went. Do you think it is more or less effective for you than note cards? Why?

▌▌▌Cheat Sheets

I do this exercise with some of my classes. I tell them I have decided, as an experiment, to let them bring in one sheet of paper to their next exam with any notes they want to make. But they can only have one piece of paper and they only have 30 minutes to make their "cheat sheet." Some classes realize I'm tricking them into studying really hard for 30 minutes, others fall for it and actually think that I'm seriously going to let them use cheat sheets. Regardless, they spend 30 minutes feverishly writing down important information. As they go through this process, think about what they are doing:

- Considering important vocabulary terms
- Deciding what formulas need to be included
- Predicting the kinds of questions that will be on the exam
- Flipping through and reviewing their notes to survey the entire body of material
- Looking for trouble spots so they can make a note that will help them
- Writing all these important things down

These are all powerful study tools. Most importantly, the process of looking at information, assessing what is most important, and then writing those things down is an extremely powerful memory technique. The things that go down on the sheet are not likely to be forgotten any time soon. So just creating the sheet is extremely helpful for memory, but on top of that, after it is made you have one sheet with important formulas, definitions, key points, and key examples.

XERCISE

Making a Cheat Sheet

This activity will have to wait until two or three days before your next test. Pretend that your professor has decided to let you bring one sheet of notes (front and back) into the exam. Really try to put yourself in the role. What would you want on the sheet? How would you organize it? Take as long as you need to create the cheat sheet. After the test, use the space below to discuss how the technique worked for you. How did you create the sheet? How did you organize it? And did creating it have a positive effect on your retention for the exam? Do you think you'll use this technique in the future? Why or why not?

We conclude this section on specific memory techniques by reiterating a warning given earlier. Make sure you do not spend so much time trying to develop these techniques that you begin to neglect doing homework in favor of them. These techniques should supplement your normal study routine.

GENERAL STUDY TIPS

We have covered a good bit in this book about how to schedule and make use of your study time. There are a couple more important ideas I would like to mention before I summarize what we have discussed.

Tutor Center

Most colleges have a tutor center where students can get help in all their subjects. In fact, most schools have a "Math Center" or a "Math and Science Center." You can go to these places for help if you have questions. A lot of students will even plan their study time so they can sit in the Math Center and do homework and ask questions as they arise. Ask classmates or other students in the center who the best tutors are and try to get their help. Most tutor centers stay busy, so have your book and notes open and ask specific questions. Keep in mind that the tutors are not your professor. For the most part, they are not there to re-teach you the material, and often when they try the results are horrible. But they can tell you where you went wrong on a problem or help you get started on one, and that can make a world of difference.

Study Groups

Students who join study groups succeed; it's that simple. Study groups provide support from other motivated students and an opportunity for you to ask questions and explain concepts. There is an old saying, "Nothing teaches like teaching." It means that when you explain a concept to someone your understanding of that concept is greatly enhanced.

If you are a motivated student, it will not be difficult to find a study group. You will naturally migrate to the students who are talking about the course and asking questions. Once you've found those students, just say, "Anybody want to get together and study?" About half of them will say yes. Study groups are very popular among students who are driven to succeed, because they know study groups work.

If you are in a class where you are having a hard time getting a study group together, go to your professor before class or during office hours. Hand the professor a piece of paper with your name, phone number, and e-mail address on it and tell him or her that you want to form a study group. Ask the professor to advertise your name and number in class. He or she will be blown away by your level of commitment and will sing your praises and the praises of study groups at the beginning of the next class. Professors love it when their students form study groups. It shows that students care and, as I have said, study groups work. . . period.

If the other students in your class don't want to join a study group, you might try making connections with other students at your school's tutor center.

Importance of a Comprehensive Approach

While homework is the most time-consuming and important part of studying, it is important you don't equate "studying" with "doing homework." Here are the study strategies we have discussed in this book:

- Doing homework effectively
- Surveying and/or reading your textbook
- Creating a vocabulary list
- Reworking and reviewing notes
- Preparing questions for your professor or a study group
- Using memory techniques like note cards, learning maps, and/or a cheat sheet
- Quizzing yourself

And while you obviously don't have to do all these things, you should balance your study time with several of them. Students who do homework exclusively tend to struggle with tests. We will discuss why in the next chapter.

 I hated every minute of training. But I said, 'Don't quit. Suffer now and live the rest of your life as a champion.'

MUHAMMAD ALI, *boxing champion*

WORD PROBLEMS

Why Word Problems Are Hard and Why Word Problems Are Easy

The difficult part of word problems is translating English into math. Most students look at a word problem in paragraph form and have a tough time thinking of how all those words are supposed to result in an equation. Believe it or not, the person who wrote your textbook understands that. And he or she wrote the textbook to try to make it easier for you. Today's textbooks are actually very student-friendly when it comes to word problems. Authors have gone out of their way to give students the tools they need to work in this tricky area.

To start getting better at word problems, go to the section in your textbook that addresses how to solve word problems. This section will be toward the front of the book and will be titled something like "Introduction to Problem Solving." In this section, the author will give you a step-by-step approach for dealing with word problems. This is very important because all the word problems in the entire book are designed to be solved by these steps. Also, the examples of word problems will follow this format. Don't take their suggested approach as idealistic or rhetorical. Take the time to familiarize yourself with it.

Another thing that helps make word problems manageable is that in most chapters there are only certain types. You may see a section on "Mixture Problems" or "Motion Problems." In reality, there is only one kind of "Mixture Problem." The problems may involve acid, money, or mixed nuts, but they all are solved the same way. It's the same with motion problems. The problems may involve two people biking, or trains, or whatever, but the process for solving them is not much different from problem to problem. Try to identify what types of word problems are in a section and focus on how those types are solved.

Note cards can be one great way to help yourself become proficient in how a certain type of problem is solved. Make two or three note cards with the same type of problem on them. Quiz yourself repeatedly on those note cards until you can do those problems easily, then see if you can do another problem fresh out of the book. If you still can't, put that problem on its own note card.

If you like the learning maps study technique you could create a learning map that has "WORD PROBLEMS FOR CHAPTER #_____" as the center title and then a bubble for each type of problem in that chapter. You can connect in examples, formulas, and tips for helping you solve problems correctly.

The last piece of advice I want to give about word problems is that too often students don't take the time to understand the process for how to solve one problem before they move to the next one. A student will get stumped on a word problem and use the back of the book, or a solutions manual, or a tutor to help solve it. Then when he sees the solution, he thinks, "I sure am glad that's over; now I can move on." NO! If you have to get help to solve a word problem, that's fine. But once you have seen how it is worked, try it on your own to make sure you have mastered the skill for yourself. Do this *before* you move to the next word problem. In my class we often do group work. Students will ask how to do a problem. I'll show them how to do it on the board and they'll say, "Okay, I get it." Then I erase the solution real quick and say, "Good, then you can do it on your own." And often they are stumped, which means they didn't really get it at all—even if they thought they did. You should keep this in mind for all types of homework problems, but it is particularly important for word problems.

10 Test Taking

INTRODUCTION

I chose to put this chapter last because in many ways it is the culmination of what we have learned. Test taking is the area in your math class where you stand to make the greatest gains. Most students who are not passing math are not passing because of their exam grades. Often, you can get a good homework grade even if you don't really understand the material at the level you should. The answers are, after all, in the back of book. But you can't fool an exam. An exam will show for sure whether or not you understand the material. It's just you, a pencil, and your brain.

Your Current Approach

In the space below, explore how you are currently preparing for exams. Discuss timetable and any special techniques you use.

_____

_____

_____

_____

_____

_____

_____

_____

The most common misconception about test taking is that doing homework alone should provide a student with the skills he or she needs to pass a test. This is an understandable expectation, but not a realistic one. Think about all the resources that are available to you when you attempt your homework:

- Your brain
- A pencil
- Your textbook
- Class notes
- A calculator
- Your friend's or family's expertise in math
- Your school's tutor center
- The back of the book!

For an exam, this list is reduced to:
- Your brain
- A pencil
- A calculator (perhaps, depending on the professor)

It is important that you don't misunderstand what I'm saying; homework is the most important thing you will do in a math class. The point I'm trying to make is that doing homework and preparing for an exam are not quite the same thing. Homework is an important part of preparing for an exam. But if other principles of test preparation are ignored, homework alone is likely to leave you deficient when the exam comes.

THE PLAY'S THE THING

The best analogy I've ever heard for the distinction between homework and test preparation is that of the theatrical play. Taking a test is like performing on opening night of a play. When you do your homework, you are using the script—that is, your book and notes. This is an important process. You can't have a play without a script. But, any actor will tell you that just studying a script does not ensure a good performance on opening night. In fact, if actors rely too much on a script it becomes a crutch and they don't really learn their lines at all. To bridge the gap between the script and the actual performance, actors use a dress rehearsal. Every actor knows that this is crucial because even when you think you have everything ready, something usually goes wrong at the dress rehearsal. But that's okay because then there's time to fix it. In fact, no actor in her right mind would go into a performance without a dress rehearsal! For most students, the "dress rehearsal" is what is missing from their exam preparation. They study their homework and book endlessly, without realizing that those materials can become a crutch. You have to put down the homework and textbook BEFORE the test to make sure you can do it on your own. Because when the curtain goes up, it's just you and your pencil.

Your Dress Rehearsal: The Practice Test

Nothing will improve your test scores more than taking a practice test. And I'll bet you have a practice test at the end of every chapter in your textbook. You should use this test to give yourself a *dress rehearsal*. A day or two before the exam, when your homework is done, you will be ready to take the practice test. Find a quiet place with no distractions, and give yourself about as much time as you'll have for the real exam. **Do not use your book or notes while you are taking a practice test.** If you do, you will be completely defeating the purpose. After you have finished the test or your time is up, go to the back of the book and grade yourself. In regular sections, the book only has the answers to the odd problems. But for the practice test it has all the answers (because they want you to be able to do this exercise). Circle the problems you missed. If you have a calculator you can get your score by calculating:

Score = (Total number correct) ÷ (Total number of problems) × 100

Now you can review the problem types you missed. Look back over the examples and notes to see what you did wrong.

Students often report that when they try taking a practice test they feel anxious or uncomfortable. If you do, that's great! Those are feelings that would normally come up as math anxiety on the day of the exam. You are working them out in advance, so that when the real test comes you will feel confident. Students also report that they thought they knew the material, but they bombed on the practice test and had to take another approach to last-minute studying. That's GREAT! It's much better than bombing the real test.

Taking a practice test can be a little challenging when you first try it. Frankly, it's easy to give up and start using the book or your notes. To get yourself used to it, start out by taking a practice test using every odd problem. This would give you two practice tests: one from the odds and one from the evens. You could try every third problem (1, 3, 6, 9, 12, . . .) or every fifth problem (1, 5, 10, 15, . . .). As you become more comfortable with quizzing yourself you can make the practice test longer.

Making Your Own Practice Test

As I mentioned, most books have sample tests at the end of the chapter. If your book does not, or if your instructor gives exams that are different from the book's practice test, then you can use your class notes and textbook to help create a practice test for yourself. Use your notes and homework to pick anywhere from 10 to 15 problems and write those problems down. As you write down the problems, be sure to reference the page and problem number from your book or notes so that after you are done taking the test you can go back and grade yourself by looking up the right answer. Use your instructor's previous tests as a guide for how many problems to include and what level of difficulty to include. Finally, consider taking the practice test to your professor to get feedback. He or she will likely be so impressed as to give you good hints on what you should be looking at for the real exam.

Here is a quick summary of how to make and take a practice test:

- Most books have sample tests at the end of the chapter.

- You can use your class notes and textbook to help you create a practice test for yourself.
  - □ Make sure to write the page and problem number so you can grade yourself.
  - □ Your instructor's previous tests can be a good guide for level of difficulty.

- Taking the test:
  - □ Give yourself an appropriate amount of time.
  - □ Eliminate distractions.
  - □ Do not use your book or notes.

- Once you are finished:
  - □ Use the back of the book to grade yourself.
  - □ Review the problem types you missed.
  - □ Take the practice test to your instructor to get feedback.

Taking a practice test or quiz is a good idea any time. Don't just use this technique when the exam comes. Make a practice quiz for individual sections in the chapter you are working on. It builds confidence and reassures you that you can do the material on your own. In my opinion, habitually taking self-quizzes and practice tests before exams is far and away the best way to deal with math anxiety.

Taking a Practice Test

This is another exercise that will have to wait until two or three days before your next exam. It is one of the most important exercises in this whole book (if not *the* most important), so make sure to give it your attention. Try taking a practice test for your next exam following the guidelines given above. In the space below, reflect on what you thought of the exercise both before and after the exam.

Before the Exam: How was your experience in taking the practice test?

_____

_____

_____

_____

_____

After the Exam: Was taking the practice test an effective method of preparation?

_____

_____

_____

_____

_____

OTHER IDEAS ON PREPARING FOR TESTS

You need to start studying for the exam the day you start the new chapter. The first step in this process is doing your homework. As mentioned before, unfortunately, a popular trend among people who struggle in math is to stop there and spend the time between then and the exam forgetting how to do the techniques they learned in homework. You need other techniques that will help you retain the information by transferring it into solid long-term memory. This can be accomplished by note cards, vocabulary lists, reviewing notes frequently, or other methods mentioned in this book.

One week before the exam, it's time to step it up a notch. If you normally review your chapter notes and homework for 15 minutes a day, increase it to 30 minutes. Also, use your class notes and the textbook to make a list of potential exam questions. Anticipate questions that you think will be on the exam. Go to your instructor and bounce some of your ideas off of him or her. Ask for suggestions on what to study. Take the problems that you have had a hard time doing, and ask if they are "fair game" for the test.

Two or three days before the exam, take a practice test. This will give you plenty of time to review the problem types that you miss. It will also give you some time to compensate and vary your approach if you do poorly on it. Make sure you take your practice test far enough in advance so that you can interact with your professor about the results.

The day before the exam, ideally you should just be doing last-minute tune-ups and not trying to cram material into your head.

Cramming

For those who don't know, "cramming" is a study technique where a student puts off preparation until the very last minute, then spends the night before the test trying to "cram" a chapter's worth of material into his or her brain. Often, people who cram will stay up almost all night (if not all night) before the test trying to learn last-minute shortcuts.

The fact that you are reading a book on how to succeed in math class is a good sign that you aren't into cramming, so I will save you an extensive finger-waving lecture on how terrible it is. Students who cram are settling for mediocrity; theirs is the mindset of an underachiever. Your brain is forced to make a lot of weak connections, and you forget everything you "learned" five minutes after the test. This makes the next chapter even harder. Cramming is a high-stress activity and not good for you personally. It reinforces horrible personal and professional habits.

If situations arise in your life that put you behind and make it necessary for you to study a lot of material in a short amount of time just before a test, then so be it. But do not think of it as cramming. You are above cramming. You manage your time wisely. You are serious about your education. Your mindset is that of an achiever. You are in the habit of preparing well in advance; you just got caught in a crunch. No problem. Next time will be different.

Special Situations

Here are some common situations that present problems with staying on top of test preparation and some ways of dealing with them.

- **Instructors who test often:** Some students have an instructor who likes to test very often. Some instructors even test once a week. This can make it difficult to stay ahead of the game and prepare well in advance. You may have to take your practice test the day before the exam. If you do, make sure you have addressed any problem spots ahead of time so that if you miss a question on the practice test you can figure out your mistake on your own.

- **Instructors who teach new material the day before the exam:** Most instructors provide class time for review before the day of the test. Some instructors, however, teach new material right up until the class period before the test. If you still want to take the practice test well in advance, skim through and try to indicate the problems that involve material you haven't covered yet and hold off on those problems until later. You can attempt them immediately after you finish your homework for those last-minute sections.

TEST DAY

Okay! Nice work. You've been diligent in applying good study habits. Now the big day is here. Here is your test-day checklist:

_____ Get a good night's sleep.

_____ Have a good breakfast that balances protein, carbohydrates, and sugar from fruit.

_____ Review all the material on the cheat sheet you have created.

_____ Briefly review your notes and any problem types that you have struggled with.

_____ Bring everything you need to the exam:

_____ Two pencils

_____ Eraser

_____ Ruler

_____ Calculator (if you're allowed to use one)

_____ Scratch paper

_____ Bottled water and perhaps coffee or a soda

_____ Hard candy *

Arrive at the exam 10 minutes early. This will ensure that nothing unexpected happens that will make you start late. It will also give you a good chance to settle into your environment. DO NOT spend these 10 minutes feverishly looking over your notes. This usually creates a state of anxiety. You have done your job. Talk to the other students who have come early. Now would be a good time to ask them about that new movie or where they are from. This will help you relax. If you don't feel like talking, then try some deep breathing and maybe use one of your success mantras or just amuse yourself with positive self-talk: "I am a mathematical candy store. I make math look like riding a tricycle. This test is my friend. Happy test." This may sound silly, but it sure beats freaking out! Speaking of which. . .

When you get to an exam early you may encounter other students who are busy stressing out about the test and trying to do last-minute studying. If their negative "vibe" starts to make you feel anxious, set your stuff on your desk so you're ready to go, then step outside and get away from the madness.

---

*Why hard candy at an exam? I have seen a few studies on the internet that claim hard candy reduces anxiety during exams. I think the idea is that your mind has something pleasant to distract it from anxious thoughts and feelings. Several of my students swear by this and always bring hard candy to their exams.

During the Exam

Test time. The tests are passed out and you're ready to show the world what your new approach to study skills has accomplished. You have done very well with your preparation, but there are certain things you should know that will make you a better test taker.

How to Take an Exam: Do the exam using these steps in this order to maximize your potential.

1. **Do a brain dump.** The second you get the test, turn to a blank page or use scratch paper to write down all the things you are afraid you might forget during the test (particularly any formulas). Don't take more than three or four minutes to do this step, but it is important. It will loosen you up and give you confidence. No, your professor will not think you're cheating.

2. **Survey the test.** Next, read through every problem on the test. Don't stop to make notes on how to do any of the problems; you won't forget before you come back. Right now, just read the whole thing. Reading through a math test is like looking at a summary of the chapter. The material will quickly start to come back to you.

3. **Do a second brain dump.** If, after reading through the test, you see there is another formula you want to put down, do it now before you get started.

4. **Do the easiest problems first.** Starting at the beginning, go through the entire test and do all the problems that you can do quickly and easily. Skip the hard problems. This is a very important approach that many students neglect. Doing the easy problems first gets your mind warmed up and builds your confidence. It also ensures that if you get bogged down in hard problems and don't have time to finish you won't miss the points from these easy questions.

5. **Now do the hard problems**—but not necessarily in numerical order. Survey these problems again and start with the one you feel you know best how to solve.

6. **Be mindful of your time.** You should be aware of when half your time is up and gauge how much of the test you have done. This will tell you if you're on the right track or need to speed up.

7. **Review the test.** You should use every available minute you are given for the test. Do not second guess yourself, but you should spend any extra time double checking your arithmetic and looking over your solutions.

When the test is over, do your best to leave the test behind and not stress over the results. You did the best you could, and now it's time to reward yourself.

After the Exam

Once you get the exam back it is very important that you don't just stuff it into your notebook and let it sit. Use the exam as a tool for greater retention and skill development. Go through the exam and pat yourself on the back as you review the problems you did correctly. Then go back and redo the problems that you missed. Feel free to get help on those problems from your professor, classmates, or a tutor. But make sure that you really understand how to do those problems for yourself. Put these corrections on a separate piece of paper, staple them to the exam, three-hole punch it all, and then file it in the Exam Section in your notebook, as we mentioned in Chapter 6: Your Class Notebook. This will make a world of difference in your ability to retain the information long term, especially when it comes time for the final exam.

Putting It All Together

Of the techniques discussed in this section, which do think will be the most effective in helping you with preparing for and taking exams? Are there techniques you don't think you will use? Why?

_____

_____

_____

_____

_____

The Final Exam

In most math classes the final exam is cumulative. That is, it covers all the material in the course. For this reason, final exams tend to be a "greatest hits" collection from all the chapter exams that were given during the semester. Luckily, you have been correcting these exams and filing them in your notebook. So you have a good resource of information to study. Find problems in the homework that are similar to these exam problems.

While these old tests are a good place to start, your professor should be your primary resource for how to study for the final. Ask the professor two important questions:

1. How many questions will be on the final?

2. What should we study to prepare?

Most professors answer these questions before they are asked. Feel free to be persistent in asking the professor for tips for the final. If you get a professor who simply says, "Study everything!", then I suggest taking the material in this order:

1. Rework your old exams.

2. Rework your old quizzes if your professor gives them.

3. Rework in-class examples from throughout the semester.

4. Review your old homework assignments.

5. Review your notes.

Make sure you start studying well ahead of time for the final. All your materials are already organized, so you can begin one day by taking out your first test and flipping through it for 10 minutes to see what you still remember. Do this with each test on the following days. As the final gets closer you'll be ready step it up a notch and start reworking the more challenging problems.

If you have taken the advice outlined in this book, the final exam will be your time to shine. You will see more than ever how these techniques have helped you learn quickly and retain information. You will feel ready for the next math course.

CONCLUSION

Here is an old Indian story that I think really makes the point I want to conclude with.

A professor was taking a long trip by ship. Every night he would give talks in his room to the uneducated sailors. One night he turned to one of the older sailors and said,

"Old man, have you studied meteorology, the science of weather?"

"No," said the old sailor, "I haven't."

"Pity," said the scholar. "You have wasted *one-quarter* of your life."

The next night he asked the same old sailor,

"Old man, have you studied biology, the science of life?"

"No," said the old sailor, "I haven't."

"Pity," said the scholar. "You have wasted *one-half* of your life."

The next night he asked the sailor again,

"Old man, have you studied geology, the science of the Earth?"

"No," said the old sailor, "I haven't."

"Pity," said the scholar. "You have wasted *three-quarters* of your life."

The next night the old sailor came running into the scholar's cabin.

"Professor! Sir!" said the sailor. "Have you studied *swimology*, the science of swimming?!"

The scholar replied, "What are you talking about?!"

"Have you studied *swimology*?! Can you *swim*?!"

"No," replied the scholar, "I can't."

With a long face the old sailor said, "Oh professor, I am afraid you have wasted your *whole* life. The ship is sinking, and only those who can swim will make the shore."

The moral is that knowledge for its own sake is not nearly as meaningful as practical knowledge. I believe this book is about how to succeed in life. I believe this book is about *swimology*. This book is about taking your education seriously, planning your time, being disciplined and deliberate in your actions, and taking control of your future. This is *swimology*. The secret to success in math is the same as the secret to success in life: Work hard and work smart. A student who understands and behaves according to these principles will go much farther than a student who gets by on "smarts" and has a knack for "working the system." Effective study skills will follow you into every college course you take. If you are diligent with them they will begin to follow you in life. And one day they will follow you to the ultimate reward for your efforts: a satisfying and meaningful career.

Take from this book what you need to succeed. Apply it with diligence, persistence, and patience. And come out successful.